Voltage and Patch Clamping With Microelectrodes

PUBLICATIONS COMMITTEE

H. E. Morgan, *Chairman*
W. F. Ganong
E. E. Windhager

S. R. Geiger, *Publications Manager and Executive Editor*
B. B. Rauner, *Production Manager*
J. E. Pedersen, L. S. Chambers, S. Mann, *Editorial Staff*
C. J. Gillespie, *Indexer*

Voltage and Patch Clamping With Microelectrodes

EDITORS
Thomas G. Smith, Jr.
Laboratory of Neurophysiology
National Institute of Neurological and
Communicative Disorders and Stroke
Bethesda, Maryland

Harold Lecar
Laboratory of Biophysics
National Institute of Neurological and
Communicative Disorders and Stroke
Bethesda, Maryland

Steven J. Redman
Experimental Neurology Unit
The John Curtin School of Medical Research
The Australian National University
Canberra, Australia

Peter W. Gage
Department of Physiology
The John Curtin School of Medical Research
The Australian National University
Canberra, Australia

AMERICAN PHYSIOLOGICAL SOCIETY
Bethesda, Maryland

© Copyright 1985, American Physiological Society

Library of Congress Catalog Card Number 84-18550

International Standard Book Number 0-683-07773-2

Printed in the United States of America by
Waverly Press, Inc., Baltimore, Maryland 21202

Distributed for the American Physiological Society by
The Williams & Wilkins Company, Baltimore, Maryland 21202

Preface

The purpose of this book is to bring under one cover a reasonably comprehensive treatment of the various methods of voltage and patch clamping with microelectrodes. These clamping techniques are among the most powerful methods available for studying the fundamental events of cell membrane excitation. The recent explosion in the use of microelectrode voltage- and patch-clamp techniques and the absence of a ready source of information about these methodologies persuaded us of the need for a book such as this. We have been fortunate in obtaining authors who are internationally recognized as experts in their respective fields, and their efforts have produced a result we believe will be useful to many.

This book is the outcome of a workshop on voltage clamping with microelectrodes organized and run by the editors of this book. The workshop was part of a satellite symposium on synaptic transmission, organized by the latter two editors of this book and held on Hayman Island off the coast of Queensland, Australia, September 4–10, 1983. The symposium was a satellite of the International Physiological Congress of the International Union of Physiological Sciences held in Sydney, Australia, August 28 through September 3, 1983.

For clarity, we wish to define the audience we seek to reach. To make our book most useful, we aimed at an audience composed of graduate students and postdoctoral fellows in biophysics, neurobiology, and electrophysiology who wish to undertake experiments requiring the use of the voltage and/or patch clamp. One purpose is to help experimenters select the best technique for their particular application because there is no universally "best" voltage- or patch-clamp system. Pursuant to this goal is some consideration of the theory and practice and the advantages and drawbacks of each type of clamp. Another purpose is to provide some "how-to" information, as well as considerations of limitations and artifacts, about each system. These and other factors may expose us to the possible criticism that we have been needlessly detailed. This is certain to be true for voltage-clamp sophisticates. But if blame is due, it belongs to the editors because we asked the authors to write their papers for scientists who are inexperienced in voltage and patch clamping. Finally, we recognize that all technical books rapidly become obsolete with the advance of technology. Nonetheless we believe this book is needed during this time of rapid expansion of the techniques of voltage and patch clamping in electrophysiological research.

In closing, we would be remiss if we did not thank a number of people who played key roles in this enterprise—the book and the workshop at the satellite symposium on Hayman Island. We are grateful to Robert Porter, Director of the John Curtin School of Medical Research, because his provocative but considered challenge provided the impetus for our enterprise and to Orr E. Reynolds, Executive Secretary-Treasurer of the American Physiological Society, because his unfailing good advice, excellent management of our affairs, and successful help in obtaining funding made the workshop possible. We thank Howard E. Morgan, Chairman of the Publications Committee of the Society, for persuading the committee of the need for this book and for the Society to be the publisher. (It should be noted that this book has itself been something of an experiment: it represents the first venture by the editors and the American Physiological Society of accepting manuscripts on computer tapes and disks. Nancy Crawford, of the National Institutes of Health in Bethesda, Maryland, has been especially helpful in this effort.) Finally, we express our genuine thanks to the authors who responded so promptly and patiently to our constant pressure and to those nameless reviewers of the chapters who have contributed measurably to any success this book may enjoy.

Thomas G. Smith, Jr.
Harold Lecar
Steven J. Redman
Peter W. Gage

Contents

1. Introduction 1
   Thomas G. Smith, Jr.
   Harold Lecar

2. Useful Circuits for Voltage Clamping With Microelectrodes . 9
   Alan S. Finkel

3. Microelectrode Shielding 25
   Frederick Sachs

4. Conventional Voltage Clamping With Two Intracellular
   Microelectrodes 47
   Alan S. Finkel
   Peter W. Gage

5. Optimal Voltage Clamping With Single Microelectrode 95
   Alan S. Finkel
   Steven J. Redman

6. High-Resolution Patch-Clamp Techniques 121
   Anthony Auerbach
   Frederick Sachs

7. Voltage Clamp and Internal Perfusion With Suction-Pipette
   Method 151
   Arthur M. Brown
   David L. Wilson
   Yasuo Tsuda

8. Microelectrode Voltage Clamp: The Cardiac Purkinje Fiber .. 171
   Robert S. Kass
   Paul B. Bennett

9. Space-Clamp Problems When Voltage Clamping Branched Neurons With Intracellular Microelectrodes 191
    Wilfrid Rall
    Idan Segev

10. Comparison of Voltage Clamps With Microelectrode and Sucrose-Gap Techniques 217
    John W. Moore

11. Voltage Clamping Small Cells 231
    Harold Lecar
    Thomas G. Smith, Jr.

Index 257

ONE

Introduction

Thomas G. Smith, Jr.
Laboratory of Neurophysiology, National Institute of Neurological and Communicative Disorders and Stroke, Bethesda, Maryland

Harold Lecar
Laboratory of Biophysics, National Institute of Neurological and Communicative Disorders and Stroke, Bethesda, Maryland

---

The voltage clamp is one of the most powerful tools in the electrophysiologist's armamentarium for studying the membrane properties of electrically and chemically excitable cells. Voltage-clamp experiments lead to the precise measurement of the membrane ionic currents underlying excitation. From these measurements it is possible to obtain exact descriptions of the ion conductances and transient conductance changes that occur in response to electrical and chemical stimuli. In the last few years, extensions of the voltage clamp concept, (e.g., noise measurement) have been used to demonstrate how permeability change is mediated by the conformational transitions of discrete ionic channels, which constitute the fundamental molecular basis for electrophysiological events.

A variety of techniques can be used to study the electrical properties of individual cells, but in terms of sensitivity, time resolution, and the isolation of conductance currents the voltage-clamp method is unsurpassed. Voltage-clamp analysis is the definitive means of characterizing the electrical and chemical machinery of excitable membranes.

The voltage-clamp method, which uses electronic feedback to control membrane potential during transient changes of membrane conductance, was originated by K. S. Cole in the late 1940s (1, 2). Because the conductance changes that occur during electrical excitation of a nerve are entirely controlled by the membrane potential and not by membrane current, the voltage clamp "tamed" the uncontrollable nerve impulse and demonstrated the true time course of conductance change. These data led to the description of nonlinear membrane conductances that accounts for the seemingly discontinuous properties of nerve excitation (9, 10).

The study of axonal membrane response to steps in membrane potential showed that the key to electrical excitability lay in the properties of time-dependent voltage-sensitive membrane conductance elements. The Hodgkin-Huxley theory (9, 10) embodied the results of the voltage-clamp studies in a complete phenomenological description of the membrane conductance changes underlying excitability. This picture has been the cornerstone of all work on electrical excitation for the last 30 years.

The evolution of the basic ideas of voltage clamp from Cole's original insights on how to tame the action potential to the synthesis of the Hodgkin-Huxley theory is described in Hodgkin's Nobel Prize address (9)

> During the summer of 1947 Cole and Marmont[41,42] developed a technique for impaling squid axons with long metallic electrodes; with the aid of electronic feedback they were then able to apply current uniformly to the membrane and to avoid the complications introduced by spread of current in a cable-like structure. Cole[41] also carried out an important type of experiment, again using feedback, in which the potential difference across the membrane is made to undergo a step-like change and the experimental variable is the current which flows through the membrane. In Cole's experiments a single internal electrode was used for recording potential and passing current; since the current may be large, electrode polarization introduces an error and makes it difficult to use steps longer than a millisecond. However, the essential features of the experiments, notably the existence of a phase of inward current over a range of depolarizations, are plainly shown in the records, which Cole[41] obtained in 1947. It was obvious that the method could be improved by inserting two internal electrodes, one for current, the other for voltage, and by employing a feedback amplifier to supply the current needed to maintain a constant voltage. Cole, Marmont and I discussed this possibility in the spring of 1948 and it was used at Plymouth the following summer by Huxley, Katz and myself[43]. Further improvements were made during the winter and in 1949 we obtained a large number of records which were analysed in Cambridge during the next two years[44].
> 41. K.S. Cole, Arch. Sci. Physiol., 3 (1949) 253.
> 42. G. Marmont, J. Cellular Comp. Physiol., 34 (1949) 351.
> 43. A.L. Hodgkin, A.F. Huxley and B. Katz, Arch. Sci. Physiol., 3 (1949) 129.
> 44. A.L. Hodgkin, A.F. Huxley and B. Katz, J. Physiol. (London), 116 (1952) 424, A.L. Hodgkin and A.F. Huxley, J. Physiol. (London), 116 (1952) 449, 473, 497, 117 (1952) 500; 122 (1953) 403.

As this statement makes apparent, the voltage-clamp technical problem is twofold: establishment of an equipotential region of membrane (space clamp) and temporal control of the membrane potential. For a long time after the original work on the squid axon each new excitable cell examined was a special case with its own considerations of cell geometry and electronic methods. Some triumphs were the potentiometric clamping of myelinated nerve and the sucrose-gap technique (see the chapter by Moore). However, the straightforward use of microelectrodes to clamp cells took some time to evolve.

Thus the purpose of the initial development and application of the voltage-clamp technique was to study the electrical properties of the action potential or spike in axons. The essential results are now historical and form basic tenets of electrophysiology and biophysics (1, 2, 9, 10). The theory and practice of voltage clamping axons have been dealt with exhaustively elsewhere and are not covered here (16, 17).

Although microelectrodes are often used in axonal voltage clamping to monitor membrane potential, such use does not fall within the scope of what is usually meant by "voltage clamping with microelectrodes." This expression is traditionally reserved for experiments in which the microelectrode is used not only to record membrane potential but also where the same or another electrode is employed to supply membrane current. In the study of giant axons the membrane current is usually delivered via an axial wire within the axon or through a cut nerve end (sucrose gap). In experiments with microelectrodes the fundamental principles are the same as in clamping axons, except that no provisions are made for obtaining spatial homogeneity. Both techniques represent applications of voltage control in a negative-feedback system.

The motivation for developing and using microelectrodes for voltage clamping is straightforward—there are many membrane conductances other than those associated with action-potential generation in axons that are of fundamental importance to cell function. These conductances are often located in somatic or dendritic membranes in which the application of axial wires is impracticable. This motivation led to the development of the electronic equipment necessary to achieve an adequate degree of membrane potential control with microelectrodes. The development proceeded simultaneously in a number of laboratories in the late 1950s and was applied to a variety of electrophysiological preparations: *Aplysia* (5), cat spinal cord (4), puffer fish (6), onchidium (7), neuromuscular junction (23, 24), and eel electroplaque (12). All of these methodologies employed two intracellular microelectrodes. Although application of voltage-clamp techniques to these preparations represented considerable achievements at the time, these clamp systems had characteristics that would be unacceptable today. For example, some were AC coupled (6, 7) with no control of DC membrane potential and thus had to be operated in a pulsed mode. Others had insufficient gain and/or frequency response to hold membrane potential constant during an action potential (23, 24) and therefore could operate only over a restricted range of membrane potential. Nonetheless these early microelectrode clamps provided important information. The first reasonably adequate voltage clamp with two microelectrodes was developed by Karl Frank at the National Institutes of Health (Bethesda, Maryland) for use with *Aplysia* neurons (5). Subsequent variations of that system have produced a bounty of information and understanding of basic membrane mechanisms.

From the early 1960s until the mid-to-late 1970s the practice of voltage clamping with microelectrodes was a pretty arcane science, limited to those laboratories with the knowledge and experience (and luck) to make the mysterious machines work. Also the individual components, such as op amps, were not very good. They tended to oscillate for no apparent reason, even with abundant feedback and bypass capacitors.

In the mid 1970s two significant events occurred. First, the basic problem of feedback circuit design was addressed in a simple and understandable way by Katz and Schwartz (12). Until then, constructing voltage-clamp amplifiers was largely a cut-and-try affair. Afterwards scientists and engineers knew what was required for success and a thousand clamps grew. Second, a commercially available and successful voltage-clamp amplifier for clamping with microelectrodes was produced by the Dagan Corporation. Stories abounded of scientists from all over the world going home from Minneapolis with (literally and figuratively) little black boxes under their arms.

Because of these developments, interest in voltage clamping and the voltage-clamp literature have flowered enormously. The results have not all been beautiful (because innocents with black boxes do not always know what they are doing), but they have been mainly positive and will continue to be so.

Microelectrodes present a problem in voltage clamping because they have high resistances that are usually not constant, particularly when large membrane currents pass through the electrode. These properties impose limitations that are not easily overcome by electronic design. These considerations are discussed at length in the chapter by Finkel and Gage. If, contrary to experience, the resistance of a microelectrode were always constant and known, one could simply subtract the voltage or IR drop across the electrode generated by the flow of current from the total applied voltage to determine membrane potential. In fact, this can be done in certain restricted cases (see refs. 3 and 8 and the chapter by Lecar and Smith). Moreover one could simply use a battery to supply the command voltage. This system would result in a very fast single-microelectrode system and eliminate the need for a lot of complicated electronics.

In the bad old days, electrophysiologists employing the voltage-clamp technique in their experiments were likely to have designed and constructed their own clamping amplifiers. They knew how to use them. The users of commercial amplifiers are not always so fortunate and have recognized a need for help. For example, over a thousand reprints were requested for a recent review devoted to voltage clamping with microelectrodes (22).

Another heartening development has been the more-or-less successful development of a variety of ingenious strategies to improve on the voltage-clamp technique, including several single-microelectrode voltage clamps and patch clamping, which are a good part of the material in this book. These methods avoid a number of the problems associated with the conventional two-microelectrode technique, but they have problems of their own. In the various descriptions of clamps, it is one of our goals to discuss the design and experimental problems as an aid to clamp users.

Although voltage clamping with microelectrodes was initially developed for two electrodes, there is a variety of excitable cells that will not admit successful penetration with two microelectrodes. Several reasons for this lack of success include cell size, connective tissue, and visual inaccessibility. However, some of these cells can be penetrated successfully with a single microelectrode, even sometimes by "flying blind." In these cases a great deal of ingenuity and imagination has gone into the development of electronics to clamp the membrane potential via a single microelectrode.

Perhaps the best known and currently most widely used of these methods is the switching system developed by Wilson (25, 26) at Duke University. Another imaginative technique is a computer-assisted interactive method devised at the Frankfurt Max Planck Institute (20). A truly novel technique involves internally perfusing cell somata while voltage clamping with (14, 15) or without (13) a single microelectrode (see the chapter by Brown et al.).

An especially interesting group of methods is based on the so-called patch-electrode technique (5, 8, 11, 18–21). A patch electrode is an extracellular electrode that is placed in intimate contact with an area (or "patch") of membrane. This system was initially intended to record only the average ionic current passing through a small area of membrane and was developed to circumvent the problem of inadequate "space" clamp in nonisopotential cells (5). Subsequently the patch electrodes, which isolate a few square micrometers of membrane, were used by Neher and Sakmann (19) to record currents flowing through single ionic channels. The realization that a variant of this system could also be used as a low-impedance access to small cells or large organelles introduced a powerful and important new methodology for studying excitable membrane properties [(18, 21); see the chapter by Lecar and Smith]. Patch clamping has also been extended to membranes excised from cells, thus allowing control of the environment bathing each side of the membrane. In this way the elementary events underlying the excitability of biological membranes can be studied with as much control as can be exercised in experiments involving measurements on synthetic bilayers.

Finally, we note that voltage clamping with microelectrodes is always associated with a degree, often large, of quantitative error much greater than that found in axonal experiments. Nonetheless, important information can and has been obtained by experimenters employing voltage clamping with microelectrodes, but experimenters should have some idea of the magnitude of quantitative errors and use appropriate caution in the conclusions reached. One aim of this book is to assist in recognizing and assessing such errors.

REFERENCES

1. Cole, K. S. Dynamic electrical characteristics of the squid giant axon. *Arch. Sci. Physiol.* 3: 253–258, 1949.
2. Cole, K. S. *Membranes, Ions and Impulses.* Berkeley: Univ. of California Press, 1972.
3. Fenwick, E. M., A. Marty, and E. Neher. A patch-clamp study of bovine chromaffin cells and of their sensitivity to acetylcholine. *J. Physiol. London* 331: 577–597, 1982.
4. Frank, K., M. G. F. Fuortes, and P. G. Nelson. Voltage clamp of motoneuron soma. *Science* 130: 38–39, 1959.
5. Frank, K., and L. Tauc. Voltage-clamp studies of molluscan neuron membrane properties. In: *The Cellular Function of Membrane Transment,* edited by J. Hoffman. Englewood Cliffs, NJ: Prentice-Hall 1964, p. 113–135.
6. Hagiwara, S., and N. Saito. Membrane potential change and membrane current in supramedullary nerve cell of puffer. *J. Neurophysiol.* 22: 204–221, 1959.
7. Hagiwara, S., and N. Saito. Voltage-current relationships in nerve cell membrane of *Onchidium verruculatum. J. Physiol. London* 148: 161–179, 1959.
8. Hamill, O. P., A. Marty, E. Neher, B. Sakmann, and R. J. Sigworth. Improved patch-clamp techniques for high-resolution current recording from cells and cell-free membrane patches. *Pfluegers Arch.* 391: 85–100, 1981.
9. Hodgkin, A. L. The ionic basis of nervous conduction. In: *Nobel Prize Lectures—Physiology or Medicine, 1963–1970.* New York: Elsevier, 1972, p. 32–48.
10. Hodgkin, A. L., A. F. Huxley, and B. Katz. Measurement of current-voltage relation in the membrane of the giant axon of *Loligo. J. Physiol. London* 116: 424–448, 1952.
11. Jackson, M. B., H. Lecar, C. E. Morris, and B. S. Wong. Single-channel current recording in excitable cells. In: *Current Methods in Cellular Neurobiology. Electrophysiological Techniques,* edited by J. L. Barker and J. F. McKelvy. New York: Wiley, 1983, vol. 2, p. 61–99.
12. Katz, G. M., and T. L. Schwartz. Temporal control of voltage-clamped membranes: an examination of principles. *J. Membr. Biol.* 17: 275–291, 1974.
13. Kostyuk, P. G., O. A. Krishtal, and Y. A. Shadhovalov. Separation of sodium and calcium currents in the somatic membrane of mollusc neruons. *J. Physiol. London* 270: 545–568, 1977.
14. Lee, K. S., N. Akaike, and A. M. Brown. Properties of internally perfused, voltage-clamped, isolated nerve cell bodies. *J. Gen. Physiol.* 71: 489–508, 1978.
15. Lee, K. S., N. Akaike and, A. M. Brown. The suction pipette method for internal perfusion and voltage clamp of small excitable cells. *J. Neurosci. Methods* 2: 51–78, 1980.
16. Moore, J. W. Voltage clamp methods. In: *Biophysics and Physiology of Excitable Membranes,* edited by W. J. Adelman, Jr. New York: Van Nostrand, 1971, p. 143–167.

17. Moore, J. W., and K. S. Cole. Voltage clamp technques. In: *Physical Techniques in Biological Research. Special Methods*, edited by W. L. Nastuk. New York: Academic, 1962, vol. 4, p. 263–321.
18. Neher, E., and H. D. Lux. Voltage clamp on *Helix pomatia* neuronal membrane; current measurement of a limited area of the soma surface. *Pfluegers Arch.* 311: 272–277, 1969.
19. Neher, E., and B. Sakmann. Single channel currents recorded from membrane of denervated frog muscle fibers. *Nature London* 260: 799–802, 1976.
20. Park, M. R., W. Leber, and M. R. Klee. Single electrode voltage clamp by iteration. *J. Neurosci. Methods* 3: 271–283, 1981.
21. Sakmann, B., and Neher, E. (editors). *Single Channel Recording.* New York: Plenum, 1983.
22. Smith, T. G., Jr., J. L. Barker, B. M. Smith, and T. R. Colburn. Voltage clamping with microelectrodes. *J. Neurosci. Methods* 3: 105–128, 1980.
23. Takeuchi, A., and N. Takeuchi. Active phase of frog's end-plate potential. *J. Neurophysiol.* 22: 395–411, 1959.
24. Takeuchi, A., and N. Takeuchi. On the permeability of the end-plate membrane during the action of transmitter. *J. Physiol. London* 154: 52–67, 1960.
25. Wachtel, R. E., and W. A. Wilson, Jr. Use of the single electrode voltage clamp to perform noise and relaxation studies of acetylcholine-activated channels in *Aplysia* neurons. *J. Neurosci. Methods* 4: 87–103, 1981.
26. Wilson, W. A., and M. M. Goldner. Voltage clamping with a single microelectrode. *J. Neurobiol.* 6: 411–422, 1975.

TWO

Useful Circuits for Voltage Clamping With Microelectrodes

Alan S. Finkel

Axon Instruments, Inc., Burlingame, California

Design of High-Speed Low-Noise Headstage: Electronic design, Minimization of input capacitance • **Design of Voltage-Controlled Current Source** • **Membrane Current Measurement:** Virtual-ground current measurement, Series current measurement • **Modifying Phase Response of Clamp Amplifier** • **Adjustment of Holding Level Before Voltage Clamping**

In this chapter the theory and use of some circuits that are generally applicable to microelectrode voltage clamps are discussed. The first section discusses the electronic and experimental requirements for achieving high-speed low-noise microelectrode voltage responses. When voltage clamping with two microelectrodes it is not necessary to apply all of the recommendations proposed in this section unless extremely fast clamp responses are needed. However, when voltage clamping with one microelectrode, fast low-noise responses are normally essential to prevent the possibility of artifacts and instabilities and to limit the amount of aliased noise, as discussed in the chapter by Finkel and Redman.

In the next section the design of a voltage-controlled current-source output stage is discussed. Again, this is more frequently applicable to single-electrode voltage clamping but can also be used as the basis for the output stage of a two-electrode voltage clamp.

Subsequent sections deal with current measurement techniques, adjustment of the phase response of the voltage-clamp amplifier, and presetting the command voltage level so that when the voltage clamp is first switched on it will automatically clamp the cell membrane at its resting value. These circuits are equally applicable to both single-electrode and two-electrode voltage clamps.

Some of the discussions use the Laplace transform method, which is described in the chapter by Finkel and Gage. All frequencies are given in rad/s.

Design of High-Speed Low-Noise Headstage

There are primary and secondary approaches to the design of a headstage, and these should be used concurrently. The former involves consideration of the physical layout of the input section and techniques for fabrication of high-speed microelectrodes. The latter, which is considered first, involves the appropriate design of electronic circuitry.

Electronic design

For a given microelectrode resistance (R_e) the recording bandwidth depends on the size of the capacitance (C_{in}) to ground between the microelectrode tip and the buffer amplifier input (Fig. 1). For analytical convenience and as a practical assumption, C_{in} is usually represented as a single lumped capacitance at the amplifier input. In practice it consists of the distributed capacitance (C1) through the wall of the immersed part of the microelectrode to the bathing solution, the capacitance (C2) through the wall of the nonimmersed part of the microelectrode and the

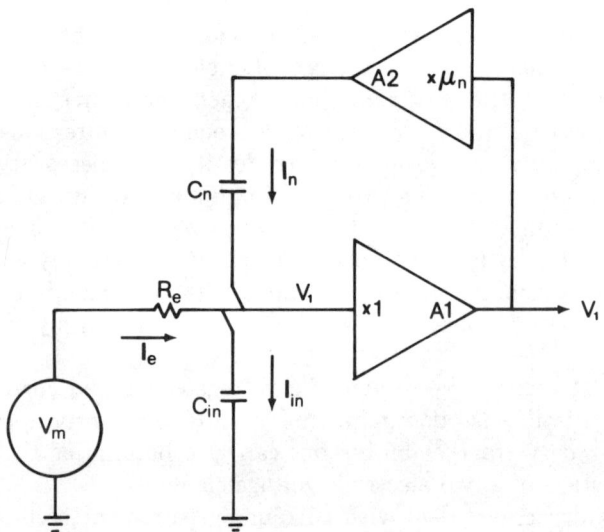

Fig. 1. Basic components and current flows in capacitance-neutralization circuit. A1, unity-gain microelectode buffer amplifier; A2, capacitance-neutralization amplifier [its gain (μ_n) is greater than +1]; V_m, membrane potential measured by microelectrode of resistance R_e; V_1, voltage recorded by A1; C_{in}, lumped equivalent of transmural capacitance of microelectrode, stray capacitance, and amplifier input capacitance; C_n, feedback capacitor. Currents are as indicated. Value of μ_n is adjusted so that $I_n = I_{in}$. Bias current of A1 is effectively zero; therefore by conservation of current into junction, I_e is zero. Until high frequencies are reached (at which point this simple analysis breaks down), presence of C_{in} is neutralized and response speed is improved.

connecting lead to the mounting hardware and other metal apparatus such as the micromanipulator, and the input capacitance (C3) of the buffer amplifier (Fig. 2). Typical values of these components are C1 = 1 pF/mm of immersed microelectrode (13), C2 = 1–5 pF depending on electrode arrangement, and C3 = 3–5 pF for a typical field-effect transistor (FET) amplifier or FET-input operational amplifier. This total C_{in} value of up to 10 pF yields an unacceptably slow response. For example, with R_e = 20 MΩ and C_{in} = 10 pF, the microelectrode time constant (τ_e = $R_e C_{in}$) is 200 μs.

To reduce τ_e the well-known technique of capacitance neutralization is commonly used. This technique is illustrated in Figure 1. Amplifier A1 is the unity-gain buffer amplifier and A2 is the capacitance-neutralization amplifier, which provides some positive feedback (gain μ_n > +1) to the input via capacitor C_n. Currents I_e, I_n, and I_{in} are as indicated in the figure. The membrane potential being measured is represented by a voltage source (V_m) at the tip of the microelectrode. The circuit topology was introduced by Bell (1) and was referred to by him as a "negative capacity" amplifier. When ideally implemented, A2 and C_n neutralize the effect of C_{in} by providing current required to charge C_{in}, which would otherwise have to be provided by current flow through R_e from V_m. That this is possible can be most easily visualized for C_n = C_{in} and μ_n = 2. In this example, the voltages across C_{in} and C_n are identical and thus I_n = I_{in}. The input current of the buffer amplifier is effectively zero, therefore by conservation of current into a node the microelectrode current (I_e)

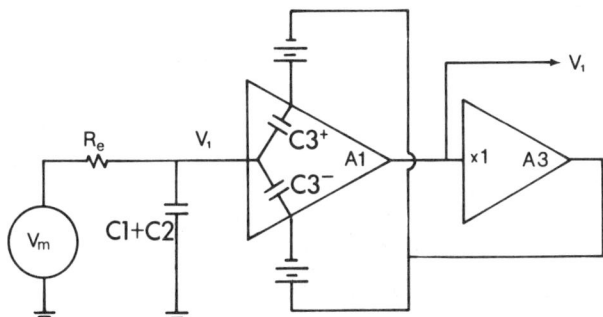

Fig. 2. Bootstrap circuit is used to eliminate dynamic effects of A1's input capacitance (C3); C3 is sum of capacitances from input to positive supply (C3⁺) and to negative supply (C3⁻). Capacitances C1 and C2 are also components of C_{in} other than C3. Amplifier A1 is powered by batteries that have input signal superimposed on them by unity-gain amplifier A3. Because voltages across C3⁺ and C3⁻ are constant, their effect on dynamic response of microelectrode is eliminated (except at very high frequencies when amplifier performance deteriorates). To ensure stability, bandwidth of A3 must be restricted to half or less of bandwidth of A1. Amplifier A3 is powered from system power supply. Batteries of A1 power supply may be real or an electronic equivalent (6).

must be zero. Thus the voltage V_1 is identical to the voltage V_m. To implement the circuit, C_n is normally fixed, and the condition $I_n = I_{in}$ is achieved by varying μ_n. For the full compliance of A1 (typically up to ±15 V) to be utilized, the compliance of A2 should be μ_n-fold greater (say up to ±30 V; $\mu_n = 2$).

This ideal analysis fails for two reasons. *1)* Amplifiers A1 and A2 have finite bandwidths and therefore the voltage across C_n is not adequately controlled at very high frequencies. The effective input capacitance (C_{eff}) that can be achieved with finite-bandwidth amplifiers is

$$C_{eff} = C_{in} + (1 - \mu_n)C_n + \tau_i/R_e \qquad (1)$$

where τ_i is the time constant of A1 and A2 together. Even if μ_n is increased to the extent that the first two terms cancel, C_{eff} remains nonzero because of the third term. The value of μ_n cannot be increased any further because the net feedback becomes positive and the circuit oscillates (see discussion of stability in the chapter by Sachs in this book). Even so, the minimum value of C_{eff} may be predicted from Equation 1 to be <0.1 pF with R_e values of 10 MΩ or more because τ_i of most operational amplifiers would be considerably smaller than 1 μs.

2) More importantly the ideal analysis fails because much of the capacitance that has been assumed to be lumped at the input is in fact distributed. Some of C1 is mixed with a portion of the microelectrode resistance and some of C3 is mixed with a portion of the substrate resistance of the input amplifier. In our experience it is not generally possible to use capacitance neutralization to reduce C_{eff} much below 10% of C_{in}, or less than ~0.3 pF, whichever is larger.[1] Because of this limitation, the use of ultrahigh-bandwidth operational amplifiers or discrete high-frequency transistors to reduce τ_i to much below 0.1 μs provides little reward.

The most convenient method for setting the optimum value of μ_n is to

---

[1] We estimate C_{eff} from $C_{eff} = \frac{1}{2}\pi f_c R_e$, where f_c is the frequency at which the amplitude of the response falls −3 dB relative to the low-frequency response. Because the limitations from the distributed nature of C_{in} are only significant at medium and high frequencies, test techniques that measure C_{eff} at low frequencies (e.g., by rate of voltage drift due to input bias current or by attenuation of sinusoid at fixed low frequency) yield low estimates of C_{eff} that are meaningless. Unreliably low estimates of C_{eff} may also arise even if a high-frequency measurement is made. For example, to measure C_{eff} caused by the amplifier input alone, it is common practice to connect a voltage source to the input via a resistor. Unless resistors with low stray capacitance across them are used, the measured value of f_c will be unreasonably large because the stray capacitance across the resistor couples a significant proportion of the high-frequency signal into the input. Small carbon-film and metal-film resistors are entirely inappropriate. Large (2–3 cm in length) thick-film resistors made from a few well-separated spirals of resistance material are most suitable. The same errors occur if f_c is measured by connecting a voltage source to a bath into which the microelectrode is immersed.

inject a current pulse (with rapidly rising leading edge) into the electrode and adjust μ_n for the fastest voltage response without overshoot. It can be shown that (if there is no capacitance across R_e) this method yields the same setting for μ_n that would be achieved by measuring the response to a voltage source at the tip (7).

Although capacitance neutralization is an essential technique for achieving the microelectrode response speeds required for single-electrode voltage clamping in particular, it has a serious drawback in that it introduces extra noise. When the bandwidth of the microelectrode recording circuit is extended, the bandwidth for measurement of the inherent microelectrode noise is also extended. However, the increase in high-frequency noise when a large amount of capacitance neutralization is used can far exceed the inherent microelectrode noise. This extra noise is due to the intrinsic input noise of A1 and A2, which is amplified by the capacitance-neutralization circuit. This noise amplification increases by 20 dB/decade at frequencies $>[R_e(C_{in} + C_n)]^{-1}$ (4, 7). The larger the values of C_{in} and C_n, the lower will be the frequency at which noise amplification begins and the greater will be the extra high-frequency noise.

Thus, to get the high-speed low-noise response required, it is necessary that primary measures to keep C_{in} (and therefore C_n) small must be taken prior to using capacitance neutralization. In addition, R_e should always be as small as possible, consistent with good cell penetrations.

Minimization of input capacitance

Several procedures to minimize C_{in} and its effects may be implemented.

1. Connect the microelectrode holder directly to the headstage input. Coaxial cable typically has capacitance between the inner conductor and the shield of ~1 pF/cm. Even a short length of cable can significantly increase C_{in} and therefore the extra high-frequency noise. Note that this is true even if the shield is connected to the output of A1 (i.e., "bootstrapped").

2. Keep the diameter of the metal headstage case near the amplifier input as large as possible and keep all other metal parts at least several centimeters away. Connect the case and other nearby metal parts to the output of A1 or A2.

3. In isolated preparations, use the minimum depth of bathing solution required to maintain the preparation viable. If practical, coat the outside of the microelectrode with a hydrophobic material to prevent a thin film of solution from creeping up the outer wall of the microelectrode. This coating may be achieved by applying a Sylgard resin (5) or by simply dipping the filled microelectrode into silicon oil (personal observation).

4. In preparations in which immersions of >1 mm must be used,

dynamic performance may be improved by shielding the microelectrode and driving the shield from the output of A1. If a wide electrode profile is tolerable, then the low-capacitance shielding technique developed by Suzuki et al. (11) would be most suitable. In their technique the voltage-recording microelectrode is slipped into a spacer electrode and the driven shield is applied to the outer surface of the spacer electrode. Because of the air space, the capacitance between the filling solution in the voltage-recording (inner) microelectrode and the shield is only one-tenth of the capacitance occurring when the shield is applied directly to the outer surface of the voltage-recording microelectrode. If a wide electrode profile is not tolerable, one of the more usual techniques described in the literature may be used with a significant loss of noise performance because of the large shield capacitance (see refs. 3, 8, 9; see also the chapter by Sachs). Normally techniques using a thin insulating layer applied by dipping or painting are not suitable for single-electrode voltage clamping because the large voltage excursions of the driven shield during current passing (up to ±15 V) may cause dielectric breakdown of the thin layer.

5. Minimize the effects of C3 by a neutralization technique. The input amplifier does not normally have a ground connection, therefore all of the input capacitance is between the input pin and the positive and negative power supplies. These components of C3 are $C3^+$ and $C3^-$, respectively, where $C3 = C3^+ + C3^-$. The effect on dynamic performance caused by this capacitance can be greatly reduced by a technique adapted to operational amplifiers by Kootsey and Johnson [(6); Fig. 2]. The input amplifier (A1) is powered by batteries. These batteries may be real or simulated by an electronic circuit (6). A unity-gain amplifier (A3) is used to drive the midpoint of the power supply batteries of A1 with the same signal recorded by A1. Because the voltages across $C3^+$ and $C3^-$ are held constant, they have almost no effect on the dynamic performance even if they are distributed in nature, although they continue to contribute to the position of the frequency at which extra noise from capacitance neutralization begins. A reduction in the total extra noise is achieved, however, because a smaller C_n may be used since C3 no longer needs to be neutralized. A further advantage occurs during large voltage swings; C3 is a poorly defined capacitance associated with the semiconductor substrate and metalization. We have observed that in some devices it is voltage dependent and contributes to signal distortion. Because the voltage across C3 is constant in the circuit shown in Figure 2, any voltage dependence is made irrelevant.

It is possible to combine the bootstrapping and capacitance neutralization of Figures 1 and 2 by giving A3 a gain greater than +1 and by applying the neutralizing current to the input via C3 (12). We have preferred not to use this technique for two reasons: *1)* any voltage dependence of C3 will increase the signal distortion and 2) if large gains must be used (because C1 and C2 are large) the power supplies of A1 may inadvertently be driven beyond the level of the input signal when the large signals that occur during current passing (as in single-electrode voltage clamp) are present. The output of A1 will cease to be related to its input, and because A1 is a part of the controlled current source (see next section) this will result in a failure to pass the desired current.

Design of Voltage-Controlled Current Source

The controlled current source (CCS) required to inject current in a single-electrode voltage clamp and in some two-electrode voltage clamps may be efficiently implemented with the technique of Colburn and Schwartz [(2); Fig. 3]. The circuit is designed to impose the command voltage (V_c) across the current-setting resistor R_o irrespective of the value of $V_e + V_m$, where V_e is the voltage across the microelectrode. To do this the gain of operational amplifier A4 must be adjusted by trimming the potentiometer RVT1 so that the gain from the input of the unity-gain buffer A1 to the output of A4 is exactly unity. The current into the microelectrode input junction is then $I_o = V_c/R_o$, which is proportional to V_c and is independent of V_m and R_e. Thus the result is a very high–impedance current source proportional only to V_c. (A negative command input, after inversion in A4, produces a positive current.) To measure the current, a differential amplifier (A5) is used to measure the net voltage across R_o. The output of A5 (V_I) is proportional to I_o. It is not always necessary to incorporate A5 because under most circumstances the voltage across R_o is already known to be equal to V_c. However, if R_e is very large or infinite (open circuit), amplifiers A1 and A4 saturate. In this case the voltage across R_o is $<V_c$ or even zero. To prevent erroneous readings during saturation, A5 should be included and the output compliance of A1 should exceed the output compliance of A4 (10).

Switch S1 is used to switch the input of the CCS from $-V_c$ for current passing to 0 V (ground) for passive voltage recording. For continuous use (e.g., two-electrode voltage clamp or bridge), S1 may be a mechanical switch operated by the experimenter. For discontinuous use (e.g., single-electrode voltage clamp or discontinuous current clamp), S1 must be a high-speed integrated-circuit switch operating at the sampling rate (see the chapter by Finkel and Redman). These switches have a finite "on" resistance (R_{on}) of 50–500 Ω. The value of R_{on} is usually poorly specified, but within the one integrated-circuit package the R_{on} values of various switch pathways are closely matched. By configuring S1 as a change-over switch, the total resistance of the input arm of A4 is kept constant,

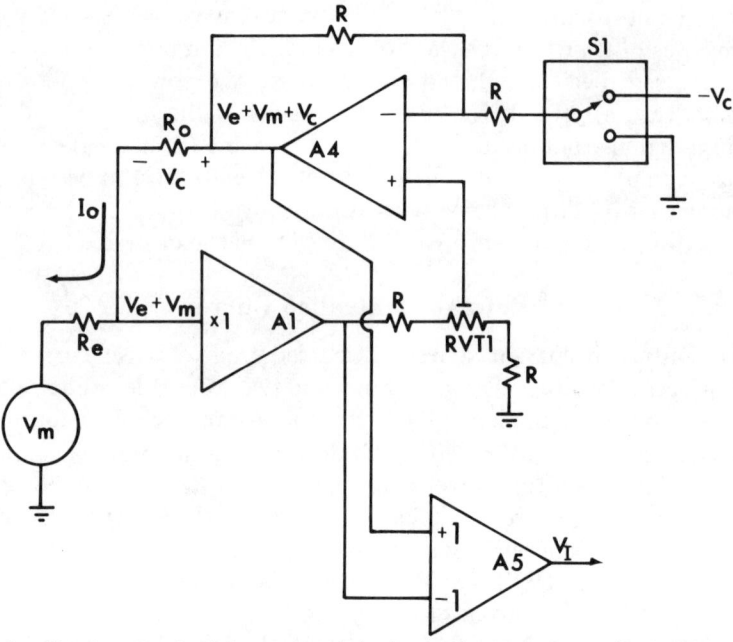

Fig. 3. Circuitry required for current injection and measurement. R_o, current-setting resistor; A4, operational amplifier; A5, differential amplifier; S1, switch; RVT1, trim pot used to ensure that gain from input of A1 to output of A4 is exactly unity (resistors are 1% or better with good long-term and temperature stability); V_c, command voltage from low-impedance source. Voltages and currents are as indicated. Switch S1 is shown in position for current passing and may be switched to ground during passive voltage recording. When S1 is in position shown, V_c is imposed across R_o irrespective of value of $V_e + V_m$. Therefore $I_o = V_c/R_o$ irrespective of value of R_e. Output (V_I) of A5 is voltage across R_o and is normally equal to V_c but may be less than that if outputs of A1 and A4 saturate.

which is necessary if the correct adjustment of RVT1 is to be the same during both periods of the cycle.

Membrane Current Measurement

There are two practical means of measuring the membrane current (I_m). Both are in common use and each suffers from disadvantages that the other does not have.

Virtual-ground current measurement

In most respects virtual-ground current measurement is the ideal technique for measuring I_m. It has one serious drawback in that, as well as measuring the membrane current, it also measures all other currents into the preparation bath. These include ionophoresis currents and line-frequency (LF) currents. The former can largely be eliminated with subtraction techniques or by floating ionophoresis current supplies with their own earth return (10). However, the latter currents are particularly difficult to prevent. They are picked up by any conducting path into the bath, including the one used for connecting the virtual ground and the fluid lines used for changing the bath solution. The LF currents are often significant in magnitude even though the voltage they generate across the low grounding impedance of the bath may be small.

The virtual-ground technique for current measurement is illustrated in Figure 4. If amplifier A6 is an ideal operational amplifier with infinite gain and zero bias current (best approximated by low-noise FET-input operational amplifiers), the potential at the negative input is constrained

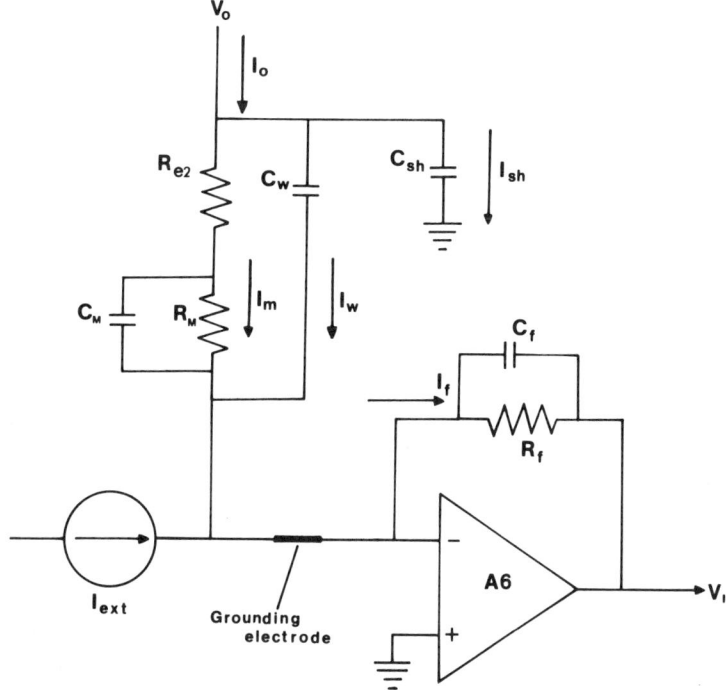

Fig. 4. Current measurement by virtual-ground method. R_{e2}, resistance of current-passing microelectrode; R_m and C_m, parallel resistance and capacitance, respectively, of cell membrane; C_w, capacitance of immersed unshielded part of wall of current-passing microelectrode; C_{sh}, capacitance due to grounded shield; A6, ideal operational amplifier (infinite gain, zero bias current); R_f and C_f, feedback network; I_{ext}, sum of all extraneous bath currents. Currents are as shown. Output of A6 (V_I) moves to keep sum of currents into inverting input equal to zero. Thus $I_f = I_m + I_w + I_{ext}$. Shield current ($I_{sh}$) is not measured. At low frequencies $V_I = -I_m R_f$. To maintain this relationship at high frequencies, C_f must be chosen so that $R_f C_f = R_{e2} C_w$.

by feedback to be at ground potential. Furthermore the current flowing into the junction at the negative input via the input network is exactly equal to the current leaving the junction via the feedback network, resistor R_f in parallel with capacitor C_f. Setting up circuit equations that satisfy these conditions and solving them with the assumption that $C_w \ll C_m$, yields the output signal

$$V_I = \frac{-(sR_{e2}C_w + 1)}{(sR_fC_f + 1)} I_m R_f \qquad (2)$$

where s is the frequency variable, C_w is the capacitance of the immersed unshielded part of the wall of the current-passing microelectrode, and R_{e2} is the resistance of the current-passing microelectrode.

By setting $R_f C_f = R_{e2} C_w$, the output voltage is exactly proportional to the membrane current; i.e., $V_I = -I_m R_f$ for all frequencies. If $R_f C_f > R_{e2} C_w$, the feedback network is overcompensated, leading to a damped response in the measurement of I_m. If $R_f C_f < R_{e2} C_w$, the feedback network is undercompensated, resulting in overshoot in the measurement of I_m. At frequencies that are low compared with $(R_f C_f)^{-1}$ and $(R_{e2} C_w)^{-1}$, Equation 2 reduces to $V_I = -I_m R_f$ even if the compensation is not perfect. Because perfect compensation depends on R_{e2}, which is variable, the best design philosophy is to minimize C_w (e.g., by using low extracellular fluid levels or a shield extending to near tip) and then to set C_f to give a slightly overcompensated response for all reasonable values of R_{e2}. Because C_w is small, quite large current-recording bandwidths can be achieved [i.e., DC to $(R_f C_f)^{-1}$].

If the virtual-ground amplifier (A6) is slow, its negative input will cease to look like a ground at high frequencies. Instead it will act as a frequency-dependent impedance in series with the cell that will adversely affect voltage-clamp performance. For this reason the gain-bandwidth product of A6 should be 1 MHz or more, and the impedance of R_f and C_f should be as low as possible (consistent with a good signal-to-noise ratio).

The extraneous bath currents are illustrated in Figure 4 by a current source (I_{ext}) that feeds straight into the summing junction of the virtual-ground circuit. The contribution to the output from these currents is

$$V_I = \frac{-1}{(sC_f R_f + 1)} I_{ext} R_f \qquad (3)$$

Thus the extraneous currents are measured to the full bandwidth limit of the virtual-ground circuit.

The current (I_{sh}) into the shield capacitance (C_{sh}) flows straight to the true ground and is therefore not measured by the virtual-ground circuit. Because I_{sh} is supplied from the low output impedance of the voltage-clamp amplifier, C_{sh} has no effect on the circuit dynamics.

Series current measurement

A technique for measuring the membrane current independently of any extraneous currents is illustrated in Figure 5. Before the output current of the clamp (I_o) flows into the current-passing microelectrode it goes through an output network in series with R_{e2}, consisting of resistor R_o in parallel with capacitor C_o. The voltage drop (V_I) across the series output network depends on I_o and is measured by an instrumentation amplifier (A5) that must have a high common-mode rejection ratio and a low input-bias current on the negative input.

Not all of I_o flows across the membrane or even into the bath; a portion flows into C_{sh} and C_w. At low and medium frequencies these currents (I_{sh} and I_w, respectively) are much less than the membrane current, thus I_o

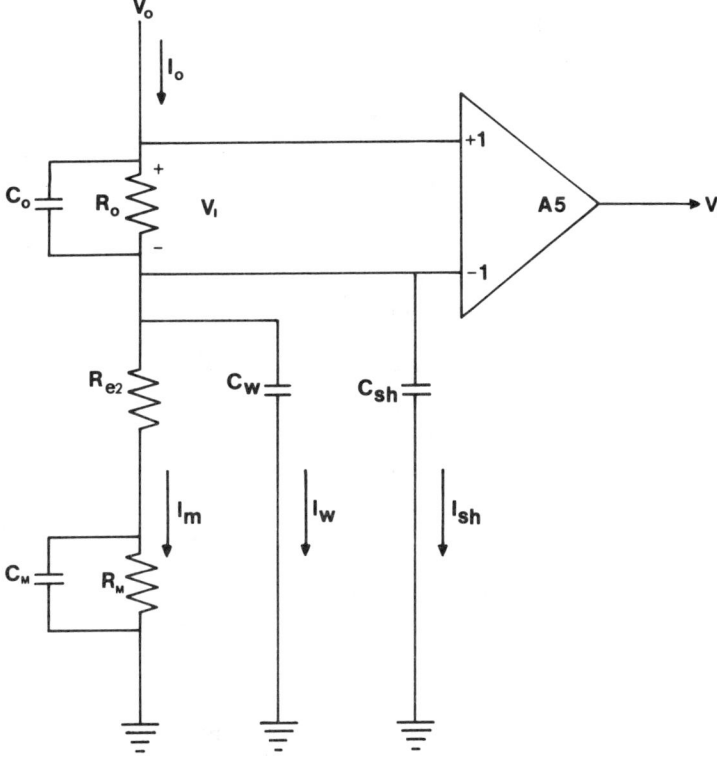

Fig. 5. Current measurement with series output network. R_o and C_o, parallel resistance and capacitance, respectively, of output network; A5, unity-gain differential amplifier. All other components and signals are as in Fig. 4. Voltage drop (V_I) across output network is proportional to I_o, where $I_o = I_m + I_w + I_{sh}$. At medium and high frequencies, I_w and I_{sh} may become significant. At low frequencies, $V_I = I_m R_o$. To maintain this relationship at medium and high frequencies, C_o must be chosen so that $R_o C_o = R_{e2}(C_{sh} + C_w)$.

$= I_m$. However, at high frequencies, $I_o = I_m + I_{sh} + I_w$. Because only I_m is of interest to the investigator, the contribution of I_{sh} and I_w to V_I constitutes a measurement artifact. To eliminate this artifact, C_o is chosen to compensate C_{sh} and C_w. Assuming that $C_{sh} + C_w \ll C_m$, the circuit can be solved to show that

$$V_I = \frac{[sR_{e2}(C_{sh} + C_w) + 1]}{(sR_oC_o + 1)} I_m R_o \qquad (4)$$

As with the previously described method for current-to-voltage conversion, the output voltage can be made linearly proportional to I_m, this time by setting $R_oC_o = R_{e2}(C_{sh} + C_w)$.

The disadvantages of the series current measurement circuit compared with the virtual-ground current measurement circuit are several. *1)* The amount of capacitance to be compensated for is typically much larger because usually $C_{sh} \gg C_w$. Thus, if the compensation is not perfect the errors may be large. *2)* The presence of an imperfectly compensated series output network loads the clamp and has an effect on the voltage-clamp step response. To minimize this, R_o should be about one third or less of the expected values of R_{e2}. *3)* The series output measurement circuit is expensive to implement when a high-voltage clamp output is used. This is because commercially available instrumentation amplifiers (e.g., Analog Devices AD521) are usually restricted to ±10-V inputs and there may be serious compliance problems. The virtual-ground circuit, however, operates independently of the output compliance.

Whereas it is practical to use a fixed value of C_f in the virtual-ground circuit, it is not practical to use a fixed value of C_o in the series current measurement circuit. This is because of the possibly large values of C_{sh}; C_o can be implemented as a variable capacitor in the headstage, but this is not generally useful because it may be necessary to adjust C_o after the microelectrode has penetrated the cell.

The same effect can be implemented by capacitance neutralization. A unity-gain buffer fitted with a capacitance-neutralization circuit is inserted in series with the negative input of the instrumentation amplifier and used to reduce the effective value of C_{sh} until the best compensation is achieved when V_o is a square voltage wave.

Modifying Phase Response of Clamp Amplifier

When voltage clamping some cell membranes, a more stable and faster response may be achieved by tailoring the phase response of V_1. Two cases in which this would be necessary are: *1)* cells with membranes that cannot be approximated by parallel resistors and capacitors (see the chapters by Rall and by Kass and Bennett) and *2)* cells with time constants that are such that a mixed fast-slow voltage clamp must be used (see the chapter by Lecar and Smith).

A simple scheme for altering the phase response of the clamp amplifier from lead to lag using only one potentiometer is shown in Figure 6. (Phase lead is achieved by placing a zero in the transfer function; phase lag is achieved by placing a pole in the transfer function.) To simplify the analysis and also to linearize the performance, the resistance of the potentiometer (RV) is made much less than the combined resistance of the resistors R1 and R2, for instance one-tenth or less. Therefore in the analysis the resistance of RV will be treated as essentially zero.

Let the fractional rotation of the wiper of the potentiometer from the lag to the lead positions be ϕ. By Laplace transform techniques the transfer function of the phase network is

$$\frac{V_j}{V_1} = \frac{R_2(1 + \phi s\, R1\, C)}{R_1 + R_2(sR_pC + 1)} \qquad (5)$$

where R_p = R1 R2/(R1 + R2) (i.e., R1 and R2 in parallel), V_j is the voltage at the junction of R1 and R2, V_1 is the input to the circuit, and C is a capacitor.

When the wiper is in the lag position, $\phi = 0$ and there is only a pole in the transfer function. Thus signal frequencies $>(R_pC)^{-1}$ are delayed by a maximum of 90° and there is high-frequency attenuation (Fig. 7,

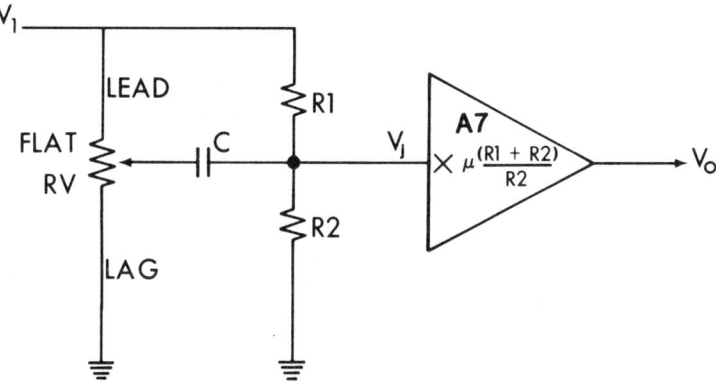

Fig. 6. Single potentiometer (RV) is used to alter phase of signal (V_1) from lag to lead. When wiper of potentiometer is in lag position, signal frequencies $>(R_pC)^{-1}$ are delayed by maximum of 90°, where R_p = R1 R2/(R1 + R2). When wiper is in lead position, signals $>(R1C)^{-1}$ are advanced by up to 90°; however, at higher frequencies [$>(R_pC)^{-1}$], phase shift is reduced back to zero. When wiper is in flat position, attenuation in potentiometer is same as attenuation in resistive divider and there is no phase shift at any frequency. To linearize transition from lag to lead, resistance of potentiometer is made 1/10th or less of resistance of resistive divider. Gain of clamp amplifier (A7) allows for overall gain μ.

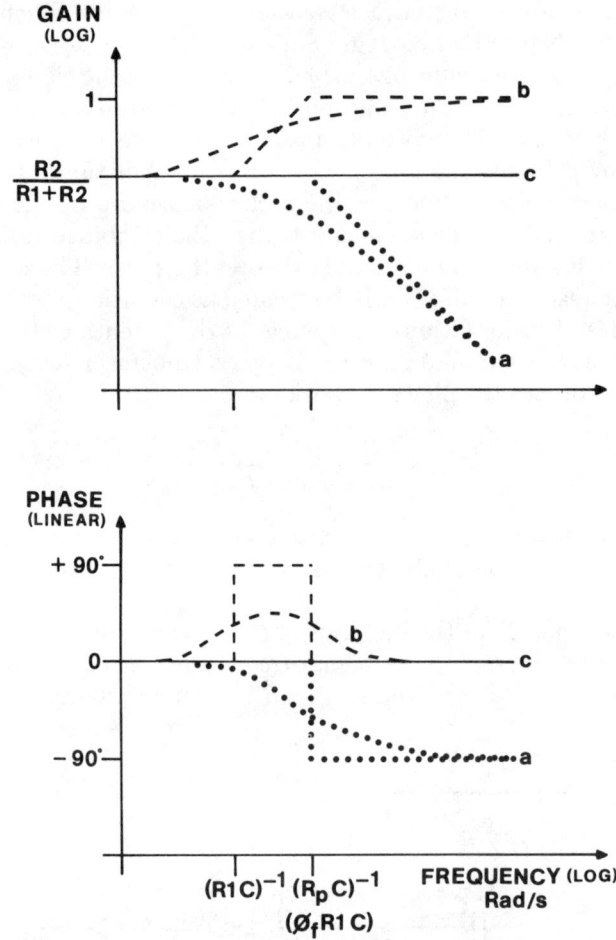

Fig. 7. Bode gain and phase plots of phase-shift network in Fig. 6. *Curves a* correspond to pure lag caused by pole at $(R_pC)^{-1}$. *Curves b* correspond to maximum lead caused by zero at $(R1C)^{-1}$, which is cancelled at high frequencies due to pole at $(R_pC)^{-1}$. *Curves c* correspond to flat response because pole and zero are coincident; i.e., $(R_pC)^{-1} = (\phi_f R1C)^{-1}$. In all cases, *straight line segments* are asymptotic approximations of smooth gain and phase plots.

line *a*). When the wiper is in the lead position, $\phi = 1$. Signals $>(R1C)^{-1}$ are phase advanced by an amount up to 90°, but at frequencies $>(R_pC)^{-1}$ the phase shift is reduced back to zero (Fig. 7, *line b*). There is a fixed increase in the high-frequency gain. When the wiper is in the flat position, $\phi = \phi_f$, the pole [at $(R_pC)^{-1}$] and the zero [now at $(\phi_f R1C)^{-1}$] cancel, and there are no phase or gain shifts (Fig. 7, *line c*). Between these extremes there is a continuum of performances.

After multiplication by the gain of A7 [$\mu(R1 + R2)/R2$] the low-frequency transfer function is μ, the desired clamp gain. The ratio of R1 to R2 is chosen to provide the required ratio of the lead to the lag frequency. To be useful, R1/R2 should be greater than or equal to unity.

Adjustment of Holding Level Before Voltage Clamping

Before switching into voltage-clamp mode it is usually desirable to set the DC level of the command potential (V_c) so that when the voltage-clamp mode is selected the cell membrane will be automatically clamped at its resting membrane potential (V_{RMP}). The required DC level of V_c depends not only on V_{RMP} but also on the total microelectrode and electronic offsets.

The simple circuit needed is illustrated in Figure 8 for a two-electrode voltage clamp. A center-zero meter is connected between the output of the voltage-recording microelectrode (ME_1) buffer and the voltage-clamp output. When the switch is open, V_c is adjusted until the meter reads zero. When the switch is closed to complete the voltage-clamp circuit,

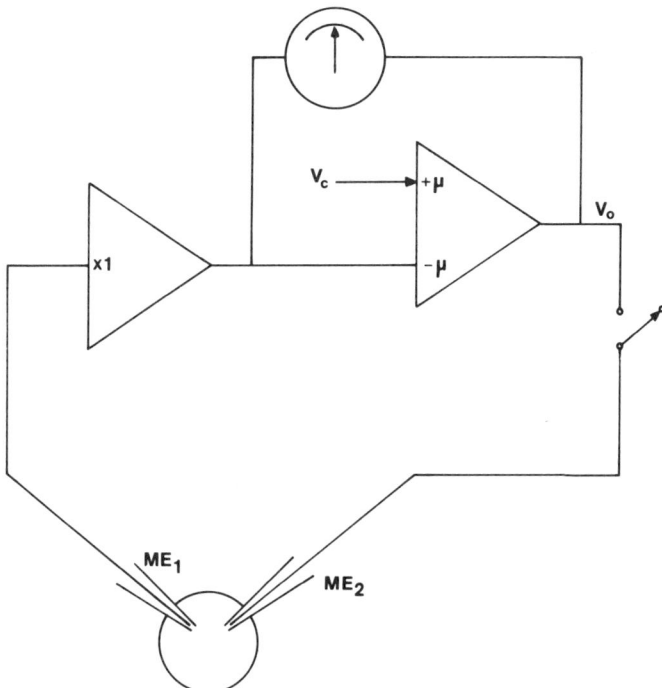

Fig. 8. Circuit to set command potential (V_c) equal to resting potential of cell before switching into voltage clamp. When meter reads zero with switch open, there will be no driving force across current-passing microelectrode (ME_2) when switch is closed.

there is no driving potential across the current-passing microelectrode (ME_2); thus the cell membrane is clamped at its resting potential. For a single-electrode voltage clamp, the meter should be connected between the output of the clamp amplifier (which connects to input of CCS) and ground.

I am grateful to Dr. S. J. Redman for his critical reading and assistance in the preparation of this paper.

REFERENCES

1. Bell, P. R. Negative-capacity amplifier. In: *Waveforms*, edited by B. Chance. New York: McGraw-Hill, 1949, p. 767–770. (MIT Radiat. Lab. vol. 19.)
2. Colburn, T. R., and E. A. Schwartz. Linear voltage control of current passed through a micropipette with variable resistance. *Mol. Biol. Eng.* 10: 504–509, 1972.
3. Finkel, A. S., and S. J. Redman. A shielded microelectrode suitable for single-electrode voltage clamping. *J. Neurosci. Methods.* In press.
4. Guld, C. Cathode follower and negative capacitance as high input impedance circuits. *Proc. IRE* 50: 1912–1927, 1962.
5. Hamill, O. P., A. Marty, E. Neher, B. Sakmann, and F. Sigworth. Improved patch-clamp techniques for high-resolution current recording from cells and cell-free membrane patches. *Pfluegers Arch.* 391: 85–100, 1981.
6. Kootsey, J. M., and E. A. Johnson. Buffer amplifier with femtofarad input capacity using operational amplifiers. *Trans. Biomed. Eng.* 20: 389–391, 1973.
7. Moore, J. W., and J. H. Gebhart. Stabilized wide-band potentiometric preamplifiers. *Proc. IRE* 50: 1928–1941, 1962.
8. Sachs, F., and R. McGarrigle. An almost completely shielded microelectrode. *J. Neurosci. Methods* 3: 151–157, 1980.
9. Schwartz, T. L., and R. H. House. A small-tipped microelectrode designed to minimize capacitance artifacts during the passage of current through the bath. *Rev. Sci. Inst.* 41: 515–517, 1970.
10. Smith, B. M., and B. J. Hoffer. A gated, high voltage iontophoresis system with accurate current monitoring. *Electroencephalog. Clin. Neurophysiol.* 44: 398–402, 1978.
11. Suzuki, K., V. Rohlicek, and E. Fromter. A quasi-totally shielded, low-capacitance glass-microelectrode with suitable amplifiers for high-frequency intracellular potential and impedance measurements. *Pfluegers Arch.* 378: 141–148, 1978.
12. Thomas, M. V. Microelectrode amplifier with improved method of input-capacitance neutralization. *Med. Biol. Eng. Comput.* 15: 450–454, 1977.
13. Woodbury, J. W. Direct membrane resting and action potential from single myelinated nerve fibres. *J. Cell. Comp. Physiol.* 39: 323–339, 1952.

THREE

Microelectrode Shielding

Frederick Sachs

*Department of Biophysical Sciences,
State University of New York, Buffalo, New York*

Appendix: Automation of Microelectrode Shielding

**James Neil
Richard McGarrigle
Frederick Sachs**

*Department of Biophysical Sciences,
State University of New York, Buffalo, New York*

Shielding for Voltage Recording • **Shielding to Decrease Interelectrode Coupling:** Effects of coupling capacity, Effects of current-electrode capacity, Effects of voltage-electrode capacity • **Shielding for Ion-Selective and Iontophoretic Electrodes** • **Resistance Requirements for Good Shields** • **Requirements for Shield Insulation** • **Shield Construction:** Reusable shields, Painted shields, Mass-production shielding • **Insulating the Shield:** Thermosetting insulation, Glass insulation, Painted insulation • **Electrode Holders** • **Summary** • **Appendix: Automation of Microelectrode Shielding:** Construction and operation, Controller operation, Detailed circuit description, Results

---

Shield (vi): to protect, defend, guard

Electrode shielding uncouples electrodes from their environment and from each other. This electrostatic isolation is essential for making accurate voltage-clamp and impedance measurements and for maintaining stability of the voltage clamp. Driven shields may also improve the accuracy of ion-selective electrodes by eliminating leakage currents across the walls of the pipette.

Shielding for Voltage Recording

Because of the high source impedance of microelectrodes, the time constant formed by the electrode resistance and its capacity limits the bandwidth. The effective capacity between an electrode and ground may be reduced by surrounding it with a conductive shield driven at the same potential as the input signal. This sort of driven shield is also known as a guard. If the shield and the electrode interior are at the same potential, there can be no charge storage and thus no capacitance. One might expect that the system response time could be made arbitrarily fast by completely shielding a microelectrode with a driven shield. Unfortunately there are inherent limitations in such a system.

To speed up microelectrode recordings, so called "negative-capacity" amplifiers have been developed that use feedback to reduce the effective input capacitance (cf. refs. 3, 5, 9, 11, 17, 20). Figure 1 shows a block diagram for a negative-capacity amplifier. The output of the first-stage follower is fed back through an amplifer with closed-loop gain A and time constant τ_a. The input capacity (C_i) includes all capacitances between the amplifier input and ground: i.e., the amplifier input capacitance (typically a few picofarads) and that part of the electrode capacitance not covered by a driven shield. The feedback capacity (C_f) includes all capacitances between the amplifier output and input: i.e. any discrete capacitors used for feedback, the capacitance of the electrode interior to the driven shield, and any input capacitance that has been "bootstrapped." Driven shields have the effect of transferring capacitance from the amplifer input to its output.

The simplest analysis of the circuit shown in Figure 1 assumes that amplifier A has a first-order response (3), and we use this assumption in this chapter. However, Shoenfield (17) has shown that the closed-loop

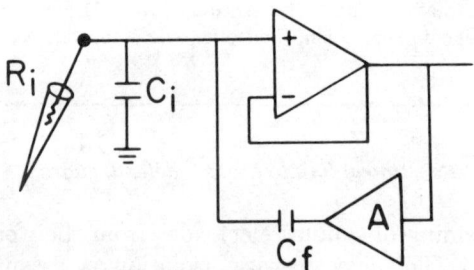

Fig. 1. Negative-capacity amplifier. C_i, all capacitance between input and ground (unshielded parts of electrode, transistor input capacity, etc.); C_f, feedback capacitance used to charge input capacity (includes all capacitance between amplifier input and output such as driven shields); R_i, resistance of microelectrode; A, gain of feedback amplifier. [From Sachs and Specht (16).]

response time may be decreased beyond that obtained with the first-order system by using amplifiers with a second-order response.

For the first-order amplifier under critically damped conditions, i.e., with the fastest nonovershooting response

$$t_r = 3.41/\omega_n \tag{1}$$

$$\omega_n = [\tau_a R_i C_i A/(A-1)]^{-0.5} \tag{2}$$

$$A = 1 + (C_i/C_f) \tag{3}$$

where t_r is the 10%–90% rise time, ω_n is the natural frequency of the complete amplifier, and R_i is the microelectrode resistance. Equations 2 and 3 have been derived assuming that $\tau_a \ll R_i C_i$. Equation 2 shows that the rise time decreases with decreasing R_i, C_i, and τ_a.

It would appear that by decreasing τ_a the amplifier can be made arbitrarily fast; however, anyone who has used a negative-capacity amplifier knows how easily they can be made to oscillate. This tendency to oscillate forms the practical limit to capacity compensation (17).

Amplifier stability can be characterized by the increase in gain required to change the system from being critically damped to oscillatory. For the negative-capacity amplifier, letting $\tau_{i,f} = R_i C_f$ and $\tau_i = R_i C_i$

$$\begin{aligned} A(\text{oscillation})/A(\text{damped}) &= 1 + 2[\tau_a/(\tau_{i,f} + \tau_i)]^{0.5} \\ &= 1 + g \end{aligned} \tag{4}$$

(Note, corresponding formula presented in Eq. 16 is incorrect. Correct version is given above.) As τ_a is decreased (faster feedback amplifier), the margin of stability (g) decreases. For example, if the input and feedback capacitances are equal and the amplifier time constant is 100 times shorter than τ_i, a 1% change in gain or a 14% change in electrode resistance will change a critically damped amplifier into an oscillator. Even in the gain range in which the amplfer does not oscillate, its response characteristics would be prone to random variations. An optimal design strategy is to minimize the rise time for a given degree of stability. Combining Equations 1–4, the shortest value of τ_a commensurate with a given stability is given by

$$\tau_a = R_i(C_i + C_f)g^2/4 \tag{5}$$

The system rise time is then given by combining Equations 1, 2, and 5

$$t_r = 0.85 g R_i(C_i + C_f) \tag{6}$$

Equation 5 shows that by constraining the stability there is no speed advantage to be gained from shielding because all capacitances removed from C_i are added to C_f. Rise time can only be reduced by decreasing R_i

and/or decreasing $C_i + C_f$. The capacitance can only be decreased by making the electrode shorter and/or moving it farther from the nearest shield or ground (20).

An unavoidable capacitance comes from the two or more picofarads of the input transistors of the follower stage. Consider an amplifier with 2 pF of input capacity and a stability of 0.01 (stable for gain variations <1%) connected to a 20-MΩ resistor. According to Equation 6, the minimal rise time is 0.34 μs. More realistically, C_i (without shielding) is at least 5 pF and the stability must be closer to 0.1 so that the minimal realistic rise time is 11 μs, which is consistent with the results of Suzuki et al. (20). If overshooting (underdamped) responses are acceptable, the rise time may be made faster, but the settling time (time until signal remains within given error band about steady-state response) will be the same or longer.

An interesting corollary of the analysis is that negative-capacity amplifiers should be designed with a bandwidth control for the feedback amplifier so that Equation 5 can be satisfied. The usual case of using a fast feedback amplifier and adjusting the gain for undercompensation can produce a slower-than-optimal response.

It should be noted that negative-capacity amplifiers are not the only way to record high-speed signals from microelectrodes. Valdiosera et al. (21) and Sachs and Specht (16) have presented microelecrode amplifiers that record electrode current rather than voltage. If the electrode resistance is known, the voltage can be calculated. Because the electrode interior is held at the potential of the summing junction, there is no change in potential and electrode capacity does not limit the response time. In fact, shielding only degrades the stability and noise performance. These current amplifiers are good for impedance studies and voltage clamps in which the command steps are <50–60 mV. For larger steps the rectifying properties of the microelectrode may degrade measurement accuracy. Current-mode amplifiers are unconditionally stable and may be made arbitrarily fast (at expense of noise). Because patch-clamp amplifiers are current amplifiers, electrode shielding only degrades system performance.

Mathias et al. (10) have presented yet another circuit to accurately measure potentials. In their scheme the electrode interior is voltage clamped to ground while feedback is used to drive the bath to the resting potential. Because the voltage electrode does not change potential, its capacity does not limit the response time and only the membrane potential is impressed across the shield insulation. This scheme avoids the accuracy problems that stem from changes in electrode resistance but makes other circuit design more difficult because the bath is not at ground (or virtual ground) and all other electrodes need to be shielded.

Shielding to Decrease Interelectrode Coupling

Microelectrode voltage clamping, with its (attempt at) high speed and its use of multiple electrodes, places stringent demands on shielding. In the two- or three-electrode clamp it is absolutely essential to isolate the voltage and current electrodes from each other. Valdiosera et al. (21) have carefully examined the effects of coupling between the current and voltage electrodes in an impedance-measuring experiment and the same general considerations apply to voltage clamping. Figure 2 is a schematic diagram of the important circuit elements of a microelectrode voltage clamp. There are three kinds of errors caused by stray capacitance. *1)* Coupling of the driving voltage (V_I) through the interelectrode or coupling capacity (C_c) produces an erroneous voltage signal. *2)* Current flow through the stray capacitance of the current electrode (C_I) bypasses the cell and thus produces an erroneous current. *3)* Capacitance between the voltage-recording electrode and the tissue bath (C_V) couples changes in the bath or local potentials into erroneous voltage signals.

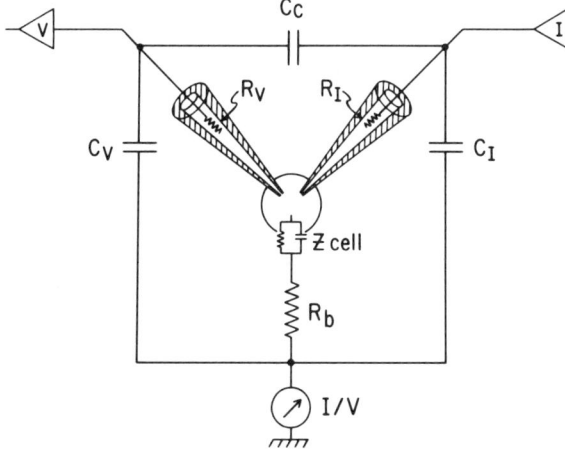

Fig. 2. Equivalent circuit of important circuit elements for microelectrode voltage-clamp or impedance measurement. R_I and R_V, resistance of current and voltage electrodes, respectively; C_I, C_V, and C_c, capacitances of current and voltage electrodes to bath and capacitance between current and voltage electrodes, respectively (for simplicity, C_I and C_V have been shown connected exclusively to bath; in practice there is also a component connected directly to ground); Z_{cell}, impedance of cell (shown as lumped impedance for simplicity); R_b, bath resistance composed of convergence resistance to cell plus reference electrode resistance plus input resistance of current-to-voltage (I/V) converter that records current; V, voltage-measuring amplifier; I, current-driving amplifier.

Effects of coupling capacity

Capacitance between the current and voltage electrodes produces a voltage signal that bypasses the cell and that may lead the cell voltage by as much as 90°. This out-of-phase signal can lead to profound errors in both impedance and voltage-clamp measurements and be a primary cause of oscillation in microelectrode voltage clamps (see the chapters by Finkel and Gage and by Lecar and Smith). Because the voltage on the stimulating electrode is generally so much larger than the cell potential and because with increasing frequency the ratio of the cell voltage to the driving voltage decreases, whereas the error current through the coupling capacity increases, a small amount of interelectrode capacity can cause significant errors. The effect of the coupling capacity depends on the resistance of both electrodes so that electronic compensation is not effective. The best solution is to shield the current electrode and perhaps the voltage electrode(s).

The magnitude of the error caused by the coupling capacity can be calculated from the circuit shown in Figure 2 as the ratio of the cell potential to the observed potential (V_c/V_o). The errors, which are predominantly in the in-phase component of the cell voltage, are clearly worst at high frequencies for which $\omega\tau_{cell} \gg 1$ (where τ_{cell} is the time constant of the cell). The calculation is simplified by assuming that 1) all other stray capacitances are negligible, 2) the voltage applied to the current electrode is much greater than the cell potential, and 3) the cell may be represented by a capacitance C_{cell} (Fig. 2). Then the maximum value of C_c commensurate with a given accuracy f ($f = 0.1$ corresponds to 90% accuracy) is

$$C_c(\omega) = C_{cell} f^{0.5}/(\omega R_v C_c)^2 \tag{7}$$

where R_v is the resistance of the voltage electrode. Consider a cell with a diameter of 50 µm, whose capacitance would be ~70 pF, and a voltage-recording microelectrode with $R_v = 20$ MΩ. To record currents with 90% accuracy at 10 kHz requires that $C_c <$ fF.

The capacity between the electrodes (C_c) can be measured approximately by placing the current electrode in recording position and the voltage electrode just over the bath. Valdiosera et al. (21) report that $C_c \sim 18$ aF with only the current electrode shielded (to within 150 µm of tip).

Even when current is applied through a microelectrode, extracellular potentials are produced by current flow through the bath resistance, which is a sum of the indifferent electrode resistance (impedance) and the convergence resistance of the cell. The first term can be reduced by using Pt-Pt-black indifferent electrodes with low impedances. Valdiosera et al. (21) used a large AC-coupled Pt-Pt-black plate in the bath in parallel with a DC-coupled reversible Ag-AgCl electrode. I have used a narrow gold-plated brass chamber coated with Pt-Pt-black and a 3 M KCl-filled pipette as the DC reference. Without attention to attaining a low-impedance bath ground, potentials in the bath can be substantial. For example, an Ag-AgCl pellet in a saline bridge has a resistance of 5–10 kΩ, which is larger than the impedance of a 1-nF cell at 10 kHz. Proper electrodes can reduce the ground resistance at audio frequencies to a few hundred ohms.

The convergence resistance of a cell is a complicated function of its geometry and the potential distribution within the cell and probably forms the ultimate limit on the accuracy of impedance or voltage-clamp measurements. A first-order correction for the effects of the convergence resistance can be made with a differential voltage measurement with the voltage electrode in recording position and then with it just outside the cell (21).

Effects of current-electrode capacity

The current-electrode capacity (C_I) shunts current directly from the driving source into the bath (cf. refs. 13 and 14). Because large voltages can appear in the driving electrode, the current through C_I can be much greater than through the cell. For frequencies at which the impedance of the cell is small with respect to the resistance of the driving electrode (R_I), the ratio of the shunt current to the cell current is given by $I_{shunt}/I_{cell} = \omega \tau_I$, where $\tau_I = R_I C_I$. For <1% of the current of a 10-MΩ electrode to leak through C_I at 10 kHz, C_I must be <16 fF, which is equivalent to an unshielded length of ~16 μm, assuming a capacitance of 1 pF/mm (22). For higher resistance electrodes, the shielding must be even more extensive. The current through C_I can also induce local fields in the bath, particularly in solutions with a high resistivity. These voltage fields add to errors produced by the current flow through the bath resistance (R_b) (21).

Effects of voltage-electrode capacity

When the bath or local potentials change, the apparent membrane potential will also change because of coupling through C_V. This error is particularly significant in preparations such as frog skin or electroplacques, where cell potentials are changed by passing current through the bath (18). The stimulation current produces large voltage gradients in the bath so that the voltage gradient across the electrode shank varies with distance along the electrode.

Shielding for Ion-Selective and Iontophoretic Electrodes

Ion-specific electrodes can have resistances in the range of 10–100 GΩ, which is comparable to the resistance of the glass of the electrode tip (12). Leakage currents through the glass can change the response characteristics of the electrode. By enclosing the electrode with an extensive driven shield, leakage currents can be reduced and the accuracy improved (7). There is no speed advantage to be gained from shielding, although speed might be increased by using current-mode amplifiers or the "funny-feedback-follower" of Mathias et al. (10).

Multibarrel iontophoretic electrodes containing a recording barrel are often used in pharmacological studies. The only way to cleanly isolate the recording barrel from the noise of the iontophoretic barrels is to use assemblies constructed from separate electrodes rather than a multibarrel blank. Sonnhof (19) has presented a multibarrel assembly in which the recording electrode is metal plated and placed within a concentric insulating barrel. Assemblies that are to be glued together can be shielded by any of the methods discussed next.

For electrodes drawn from bundles of individual capillaries (cf. ref. 1), it may be possible to partially shield the assembly by evaporating or sputtering a noble metal onto the blanks and then drawing the bundle of coated capillaries as a unit. I have found that the metal coat tends to break down a few millimeters past the shoulder during drawing because of differences in ductility between the glass and the metal, but high-quality shielding remains on the shank. The bundle would have to be insulated from the bath as a unit and the entire shielded bundle driven as a guard on the recording barrel or grounded if a current-mode amplifier is being used. It is possible to provide some shielding of a multibarrel assembly after pulling by coating the entire assembly with a conductor; however, because of the unshielded interface between the interior aspects of each barrel, the shielding will not be highly effective. Note that providing a driven shield on the recording barrel increases the capacitance of the ionophoretic barrels and will slow the delivery of drugs applied with a constant current source.

Resistance Requirements for Good Shields

A good shield should have a resistance that is low enough to reduce capacitively coupled voltages in the shield to negligible proportions. To estimate currents in the shield, one must first examine whether the electrode interior is isopotential. The electrical equivalent circuit for a shielded microelectrode is an *RC* cable with the filling solution forming the core conductor and the interior-to-shield capacitance forming the distributed capacity. At most frequencies of interest the glass conductance is negligible with respect to the admittance of the capacity. The space constant (λ) for an infinite cylindrical RC cable with no shunt conductance is given by King (4)

$$\lambda = 1/(RC\omega)^{0.5}$$
$$= r(\rho_i/C\omega)^{0.5} \qquad (8)$$

where r is the radius of the saline core, ρ_i is the resistivity of the filling solution, R is the resistance per unit length of the electrode interior, C is the capacitance per unit length between the interior and the shield, and ω is the frequency (rad/s). For a 1 M KCl filling solution, ρ is ~10 Ωcm, and if $C = 1$ pF/mm, $\lambda = 1.77 \times 10^5 r\omega^{-0.5}$ cm. Near the tip, $r = 0.5$ μm so that $\lambda > 8.9\omega^{-0.5}$ cm. Therefore, even in the tip region at the highest frequencies (10 kHz), the space constant is >350 μm, much longer than the narrow region of the tip. Thus the electrode interior may be considered isopotential, in agreement with the lack of frequency dependence of the current passed through shielded electrodes (21). With high-conductance filling solutions such as molar salts, there is no need to consider distributed resistances of a microelectrode.

By analogy with the calculation above, if the resistance per unit length of shield is significantly lower than that of the electrolyte, the shield will also be isopotential. Assuming a shield with a resistivity of ρ_s and a thickness of d and a filling solution with a resistivity of ρ_i, the ratio of the shield resistance to the core resistance per unit length (neglecting thickness of glass) is

$$R_s/R_i = r\rho_s/(2\rho_i d) \qquad (9)$$

The worst situation arises for large values of r. If a 2-mm-diam electrode is shielded with a 1-nm-thick layer of silver ($\rho_s = 1.6 \times 10^{-6}$ Ωcm) and the electrode is filled with a solution for which $\rho_i = 10$ Ωcm, then $R_s/R_i < 0.08$. Thicker coatings would decrease this value further. However, for other substances such as graphite (2), which have higher resistivities ($\rho_s = 1,400 \times 10^{-6}$ Ωcm for graphite), the shield may not be adequate for demanding applications. The shield itself must be grounded through a low-resistance connection to properly shunt the shield current. The magnitude of the shield current is given by $I = C(dV/dt)$, where C is the total capacity between the interior and the shield and dV/dt is the rate of change of the applied potential. In an extreme but not unrealistic case, a 100-V/s step applied to an electrode shielded for 5 cm will produce a shield current of 0.5 mA. To keep voltages on the shield below 1 mV, the shield must be grounded with a resistance <2 Ω.

Requirements for Shield Insulation

Generally shields must be insulated from the bath by an impedance much larger than the impedances being measured. When current is measured with a current-to-voltage (I/V) converter connected to the bath, a current leak between a grounded shield and the bath will cause a drifting DC error in the recorded current. A current leak between a driven shield and the bath can produce arbitrarily large dynamic errors proportional to the shield potential.

Driven shields pass capacity currents into the bath through the insulation-bath capacity. The magnitude of this capacity current depends primarily on the thickness of the insulation. Considering the shield-insulation-bath system as a coaxial capacitor, the capacitance is given by

$$C_{\text{coax}} = 0.056\kappa/\ln(r_1/r_2) \text{ pF/mm} \qquad (10)$$

where κ is the dielectric constant (2–4 for most insulators) and r_1 and r_2 represent the outer and inner radii of the shield insulation, respectively. The highest capacities arise when the two radii are comparable, i.e., in the larger diameter sections of the pipette. For an electrode with a nominal 1-mm radius coated with a 100-μm-thick coat of insulator with a dielectric constant of 4, the shield-bath capacitance would be 2 pF/mm. This amount of capacitance can carry substantial currents when the shield is driven at high frequencies. For the shield just considered, a 1-V, 10-kHz signal on the shield will pass 120 nA into the bath per millimeter of electrode insertion. This current would be added to the transmembrane current recorded by an I/V converter connected to the bath. When driven shields are used on current-passing electrodes, the current should be monitored at the input to the electrodes rather than from the bath. A better arrangement, however, is to ground shields on current electrodes.

The recording arrangement can make a great deal of difference in the choice of insulating techniques. If the current electrode is shielded and the current is recorded with a virtual-ground I/V converter, potentials across the shield insulation are on the order of only a few millivolts. Insulation for driven shields on voltage electrodes must be able to withstand potentials of several hundred millivolts. Insulation for driven shields on current electrodes must be able to withstand hundreds of volts.

Shield Construction

One of the most important requirements of a good shielding technique is the ease of construction because experimental success is an inverse exponential function of the number of steps in the preparation. An ideal shielding technique would be to have pipette blanks with a concentric insulated shield that could be drawn like normal microelectrodes. We are yet awaiting such a glass. However, there are techniques that minimize the handling of individual electrodes.

Reusable shields

The simplest shields enclose an ordinary microelectrode in a conductive housing, such as the common "spring" system (Fig. 3). These shields do not penetrate the bath and thus are only good for low-frequency nondemanding applications. A similar but more compact shield can be made by wrapping a small piece of aluminum foil around the bare glass electrode.

Painted shields

For high-quality shields the shielding must be applied closely to the glass so that it can extend close to the tip of the electrode. Painting electrodes with conductive paints is the most common method for shielding; however, it requires care to shield close to the tip of the electrode, and often the large grain size of the paint interferes with visibility. Nonetheless the technique requires little in the way of equipment and the materials are commonly available. Table 1 lists several brushable coatings that have been used.

The bake-on coatings adhere to the glass better than the silver-loaded paints. The organometallic coatings have the advantage of a much finer grain size than the silver-loaded paints. The organic-gold coating used by Sachs (13, 14) requires that the electrode blanks be heated to the softening point of the glass in an oxidizing (air) atmosphere, but in return the coating is tightly bonded to the glass and has a negligible grain size. The primary disadvantage of painting is the care required to shield close to the tip. Llinas et al. (8) claim coatings within 20 μm of the tip, but more common values are 50–150 μm from the tip.

Mass-production shielding

It is a distinct advantage to be able to shield electrodes in bulk so that they can be prepared in advance of the experiment. The central problem with bulk shielding is preventing the shield from covering the electrode tip. There are three published techniques for bulk coating.

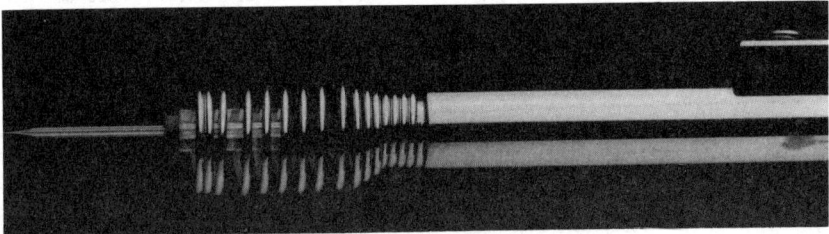

Fig. 3. Spring-type shield that is covering just electrode holder. (Courtesy WPI, Inc.)

Table 1
Brushable conductive coatings

	Drying Method	Source	Ref.
Silver paint	air dry	#21-2	
		G. C. Electronics	
		Rockford, IL	18
		SPI Supplies, high purity	
		Structure Probe Inc.	
		P.O. Box 342	
		Westchester, PA	*
	mild bake	Auromal 37M	
		Doduc KG	
		7530 Pforzheim, FRG	20
Colloidal silver	air dry	Electrodag 415 or 416	
		Acheson Colloids	
		Port Huron, MI	8
Organic silver	volatile resin	Englehard Industries	
		Electrometallics Division	
		Newark, NJ	21
Organic gold, Pt	bake on	Englehard Industries	
		Liquid Brite Gold Division	
		Newark, NJ	14, 15

* R. Mathias, personal communication.

1. Engberg et al. (2) have used a graphite-loaded spray (Graphit 33, Kontakt Chemie, Rastatt, FRG) to cover electrodes with a conductive coating. The coating is then removed from the tip by immersion in ethanol. Unfortunately ethanol has a positive meniscus; thus the coating is dissolved up to a millimeter or two from the tip. Also, the resistivity of graphite may be too high for good shielding.

2. Kottra and Frömter [(6); see also ref. 20] presented a second mass-production approach. They tried to minimize capacitance between the electrode and the shield to increase recording speed. They placed an ordinary microelectrode within a larger diameter pipette that had been internally coated with silver (Fig. 4). Because the shield was spaced away from the electrode, the capacity between the two was reduced from a nominal 1 pF/mm to 0.16 pF/mm (see Eq. 10). The internally shielded insulating jackets were produced in bulk. The outer jackets were drawn from standard stock but with a larger diameter tip than the microelectrode. To provide better insulation for the shields in the later stages of construction, the blanks were made hydrophobic by exposing them to silane vapor (dichlorodimethyl silane) for 10 s. The treated blanks were baked for 60 min at 130°C. The jackets were then mounted on a holder so that the shank-end openings were focused at a point, and the holder was placed in a vacuum evaporator with a silver bearing filament at the focus.

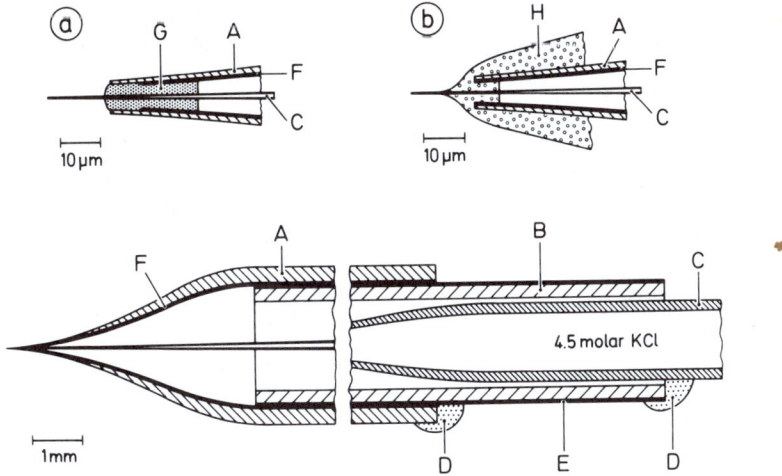

Fig. 4. Low-capacitance shielded microelectrode. Glass jacket (A) coated with silver (F) on inside; connecting capillary (B) coated with silver (E) on outside and glued to jacket with dab of epoxy (D); C, microelectrode. *Insert a* shows tip insulated with Sylgard (G); *insert b* shows tip insulated with polystyrene lacquer (H). [From Kottra and Frömter (6).]

After silver coating the interior of the jackets, the recording electrodes were carefully slid inside and glued in place. The authors claim shielding to within 10 μm of the tip. The glass of the covering pipette serves as the primary insulation. Except for the need to place each electrode within its shield under microscopic examination, the technique appears fairly simple. Electroless deposition of silver using the standard silver-mirror process might make the technique available to those without access to a vacuum evaporator.

3. The third mass-production technique coats the entire electrode with silver and then selectively removes it from the tip (15). Because this is the only batch technique with which I am personally familiar, I will treat it in a little more detail than the others. Standard microelectrodes in batches of 20–50 are placed tip up in a drilled aluminum block in either a vacuum evaporator or a sputter coater. Vacuum evaporation (few hundred milligrams of silver at distance of ~5 in) produces a thicker coating than sputtering, but sputtering is much faster and less messy. (Sputtered coating is sufficiently thin that it may be difficult to establish reliable contact with electrode holder. To provide more robust contact surface, either brush conductive paint or spray graphite on shank end of electrode.) If electrodes are to be stored before further preparation, they should be stored under vacuum to avoid oxidation of the silver, which interferes with the later stage of silver removal.

To remove silver, the pipette tips are dipped for a few micrometers into a pool of clean mercury that dissolves the silver. Because the glass-mercury meniscus is negative, there is fine control of the amount of silver removed. With an automated dipper (see **Appendix: Automation of Microelectrode Shielding**, p. 42) the removal of silver can be consistently limited to <10 μm without microscopic manipulation of any kind. Manual dipping of the tip into mercury provides less consistent results, but it is simple to shield to within 50 μm of the tip. After removing silver from the tip, the electrodes are insulated as discussed below. Note, the preferred insulation technique that uses a thermosetting wax also requires no microscopic manipulation.

Insulating the Shield

Insulation techniques depend strongly on the voltages present between the shield and the bath.

Thermosetting insulation

The simplest insulating technique adequate for grounded shields and driven shields on voltage electrodes involves dipping the tip of the electrode into a thermosetting hydrophobic wax (15). Shielded electrodes are first filled with saline and then the tips are briefly dipped into a molten wax. The electrode is held tip up for a few seconds until the wax hardens. As the wax cools, surface tension draws excess wax from the tip leaving a coating that is thin near the tip and thick further up the shaft. The breakdown resistance is adequate to withstand potentials of several hundred millivolts. When testing insulation resistance it is important to use an ohmmeter that will not apply more than a minimum voltage. Otherwise the insulation is likely to break down during the test. A simple voltage-limited go/no-go insulation tester is shown in Figure 5.

The key to making the wax technique work properly is finding the proper wax. The melting point should be below 100°C to avoid boiling the filling solution and possibly introducing bubbles. The viscosity should be such that the wax is thin enough to draw back from the tip when cooling and thick enough to provide an adequate breakdown resistance. Pyseal^R cement (#C-228, Fisher Scientific, Pittsburg, PA) seems to work well. Schwartz and House (18) recommend equal parts of ceresine and beeswax. The cement is kept fluid in a small beaker on a hotplate with the temperature adjusted for proper viscosity. Somewhat surprisingly, as long as the electrodes are filled they rarely clog during insulating. There is some evidence that the seal between the wax and the pipette may decrease its resistance after 15–20 min in saline, but the amount of decrease seems small for intracellular microelectrode applications.

Glass insulation

Several authors (15, 18, 20) have used insulating jackets made of steeply tapering large-tipped pipettes. These jackets should have a high-

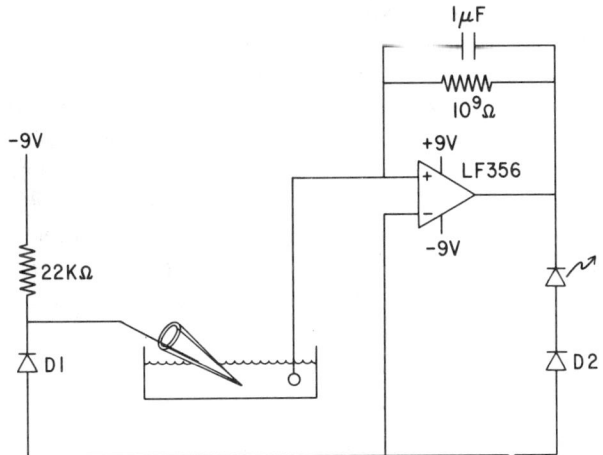

Fig. 5. Go/no-go insulation tester with limited test voltage. Silicon diode (D1) serves as 0.7-V voltage source (Ge diode could produce 0.3 V). This voltage is applied to electrode shield, and current flowing through insulation to saline bath is measured by LF356 FET input amplifier (National Semiconductor, Santa Clara, CA) configured as 1 nA/V current-to-voltage converter. The 1-nF capacitor in feedback loop is used to suppress power-line interference. Output voltage is sensed by light-emitting diode in series with silcon diode (D2) that serves to decrease sensitivity; LED will brighten if insulation resistance is less than ~10 GΩ. Tester runs on two 9-V transistor radio batteries.

enough breakdown resistance to be usable as driven shields for current electrodes. The jackets are drawn from glass with an inside diameter just sufficient to pass the outside diameter of the electrode. The jacket is made with a steep taper (by using low filament temperature on puller) so that the shoulder of the electrode does not hit the shoulder of the jacket. The technique is much simpler if both electrodes are made with a rotating puller so that they are both accurately concentric; otherwise the electrode tip may break while it is being slid inside the jacket. Placing one electrode inside the other needs to be done under a microscope with the inner electrode held in a micromanipulator so that it can be placed accurately and held with the proper amount of tip exposure during gluing.

After the electrode is in position, a dab of sticky wax or fast-drying epoxy applied to the barrel can be used to glue the electrode in place within the jacket. The jacket must then be sealed to the electrode at the tip. This seal can be made with thermosetting or drying lacquers and glues. Schwartz and House (18) recommend dipping the tip into a molten mixture of equal parts ceresine and white beeswax, which is similar to Pyseal[R] cement.

Suzuki et al. (20) glued the insulating jacket to the electrode tip with Aremcobond 517 (Solak-Chemie, 7000 Stuttgart, FRG) and then baked the electrodes for 30 min at 145°C in a ventilated oven. After baking, the tips were further covered with a lacquer (Gotekplast^R, Lurgi, 6000, Frankfurt, FRG) and dried for 10 h. The electrodes were filled with saline just before use. Kottra and Frömter (6) seem to prefer insulating the tips either by dipping them repeatedly into polystyrene lacquer (AI Speciallack, farblos, from E. Diegel, Alsfeld, BRD) and then using a hot filament to melt the lacquer back from the tip or by using Sylgard^R 185 (Dow-Corning, Midland, MI). The silicone-rubber technique is reportedly only applicable to those jackets that have been so effectively treated with a silanizing treatment (see **Shield Construction**, p. 34) that no silver has been deposited on the last 5–10 μm of the jacket tip.

Painted insulation

Although tedious, shields can be insulated by painting them with a variety of lacquers and varnishes. Polystyrene lacquers such as nail polish and "Q-dope" or varnishes such as Glyptal^R #10-9002 (G. C. Electronics, Rockford, IL and most electronic supply houses) can be brushed on over a shield. Lewis and Wills (7) use a coating of nail hardener (Cutex) followed by a coating of M-coat D^R (Micro Measurements, Raleigh, NC). Llinas et al. (8) have used several coats of an air-drying compound called Humiseal^R (Humiseal Division, Columbia Chase Corporation, Woodside, NY). The catalysed silicone rubbers [Sylgard^R 184 or 185 (Dow Corning, Midland, MI and electronic supply houses)] have excellent insulating properties. They can be brushed on and then quickly cured over a hot-air gun or left to cure overnight. However, if painted too close to the tip of an unfilled electrode, Sylgard may make the interior too hydrophobic to fill. As with all coatings, there is always a danger of clogging; however, filling the electrode prior to insulating may avoid this problem. Dipping the electrode tip in an appropriate solvent can remove clogs. However, the use of solvents is hampered if they also soften the binder used in the conductive paint of a painted shield.

Electrode Holders

To preserve the integrity of any shield, the electrode must be clamped in a shielded holder. The design criteria are that *1*) the shield be essentially complete, *2*) the electrode not be short-circuited by saline leaking from the back of the electrode into the body of the holder, and *3*) electrode installation be simple to avoid breakage. I have used a holder built like a long thin alligator clip with grooves down the middle of each jaw to match the outer diameter of the electrode. The connecting wire is shielded and, beginning near the holder, is covered with a piece of small-diameter silicone-rubber tubing. The tubing forms a tight fit over the back of the electrode and prevents the filling solution from contacting the holder.

Summary

Despite the outrageous combination of high impedance and wide bandwidth in microelectrode voltage clamps, careful attention to shielding makes it possible to obtain accurate results at frequencies in excess of 10 kHz.

REFERENCES

1. Carette, B. A new method of manufacturing multi-barreled micropipettes with projecting recording barrel. *Electroenceph. Clin. Neurophysiol.* 44: 248–250, 1978.
2. Engberg, I., J. A. Flatman, and J. D. C. Lambert. A simple and cheap method of screening glass microelectrodes (Abstract). *Brit. J. Pharmacol.* 55: 312P–313P, 1975.
3. Guld, C. Cathode follower and negative capacitance as high input impedance circuits. *Proc. IRE* 50: 1912–1927, 1962.
4. King, R. W. P. *Transmission Line Theory*. New York: Dover, 1965.
5. Kootsey, M., and E. A. Johnson. Buffer amplifier with femtofarad input capacity using operational amplifiers. *IEEE Trans. Biomed. Eng.* 20: 389–391, 1973.
6. Kottra, G., and E. Frömter. A simple method for constructing shielded, low-capacitance glass microelectrodes. *Pfluegers Arch.* 395: 156–158, 1982.
7. Lewis, S. A., and N. K. Wills. Resistive artifacts in liquid-ion exchanger microelectrode estimates of Na^+ activity in epithelial cells. *Biophys. J.* 31: 127–138, 1980.
8. Llinas, R., I. Z. Steinberg, and K. Walton. Presynaptic calcium currents in squid giant synapse. *Biophys. J.* 33: 289–321, 1981.
9. MacNichol, E. F. Negative impedance electrometer amplifiers—introduction. *Proc. IRE* 50: 1909–1911, 1962.
10. Mathias, R. T., J. L. Rae, and R. L. Eisenberg. The lens as a nonuniform spherical syncytium. *Biophys. J.* 34: 61–83, 1981.
11. Moore, J. W., and J. H. Gebhart. Stabilized wide band potentiometric preamplifiers. *Proc. IRE* 50: 1928–1941, 1962.
12. Okada, Y., and A. Inouye. Studies on the origin of the tip potential of glass microelectrode. *Biophys. Struct. Mech.* 2: 31–42, 1976.
13. Sachs, F. Electrophysiological Properties of Tissue Cultured Heart Cells Grown in a Linear Array. Syracuse, NY: Upstate Medical Center, 1970. PhD thesis.
14. Sachs, F. Electrophysiological properties of tissue cultured heart cells grown in a linear array. *J. Membr. Biol.* 28: 373–399, 1976.
15. Sachs, F., and R. McGarrigle. An almost completely shielded microelectrode. *J. Neurosci. Meth.* 3: 151–157, 1980.
16. Sachs, F., and P. Specht. Fast microelectrode headstage for voltage clamp. *Med. Biol. Eng. Comput.* 19: 316–320, 1981.
17. Schoenfeld, R. L. Bandwidth limits for neutralized input capacity amplifiers. *Proc. IRE* 50: 1942–1950, 1962.
18. Schwartz, T. L., and C. R. House. A small-tipped microelectrode designed to minimize capacitive artifacts during the passage of current through the bath. *Rev. Sci. Inst.* 41: 515–517, 1970.
19. Sonnhof, U. A multi-barreled coaxial electrode for iontophoresis and intracellular recording with a gold shield of the central pipette for capacitance neutralization. *Pfluegers Arch.* 341: 351–358, 1973.

20. Suzuki, K., V. Rohleček, and E. Frömter. A quasi-totally shielded, low capacitance glass microelectrode with suitable amplifiers for high-frequency intracellular potential and impedance measurements. *Pfluegers Arch.* 378: 141–148, 1978.
21. Valdiosera, R., C. Clausen, and R. S. Eisenberg. Measurement of the impedance of frog skeletal muscle fibers. *Biophys. J.* 14: 295–315, 1974.
22. Woodbury, J. W. Direct membrane resting and action potential from single myelinated nerve fibers. *J. Cell. Comp. Physiol.* 39: 323–339, 1952.

Appendix: Automation of Microelectrode Shielding

Sachs and McGarrigle (15) published a method for shielding microelectrodes with a film of silver that could be extended to within tens of micrometers of the tip. This appendix presents a device that will automatically and reproducibly shield microelectrodes to within 10 μm of the tip, yielding a stray capacity of <10 fF.

The shielding operation described by Sachs and McGarrigle (15) has three major steps: *1)* covering glass microelectrodes with silver (or gold) by thermal evaporation or sputtering, *2)* dissolving metal from the tip region by immersion in mercury, and *3)* insulating the shielding with a thermoplastic resin. We have developed a device that automatically accomplishes the second step.

Construction and operation

A coated electrode is attached to a vertical slide driven by a motorized micrometer (Ardel Kinematic, College Point, NY). A controller advances the pipette until it makes electrical contact with a mercury bath below. The controller waits a few seconds to ensure dissolution of the shield and then reverses. The electrode is then ready to be filled and insulated.

The mechanical portion of the device is shown in Figure A1. The mercury bath is made of steel and is grounded. The electrode holder is mounted on a Lucite support to electrically isolate it from ground. The electrode holder itself is removable to simplify placement of electrodes and is electrically connected to the resistance sense circuit in the controller.

In operation, the device proceeds sequentially through five states. State 0 is the end of cycle condition. In state 1, after the start command, the micrometer is lowered at high speed to bring the tip within a reasonable distance from the mercury. The duration of this rapid advance is set (by switch S3 in Fig. A2) in 5-s increments and is determined by the length of the microelectrode being processed and by the desire for minimum cycle time.

State 2 provides slow-speed operation for controlled depth of penetration into the mercury. Motor speed is precisely controlled because depth of penetration depends on inertial response of the motorized micrometer. The speed in this state is set by potentiometer R1.

State 3 follows contact with the mercury. The micrometer is stopped for a few seconds to ensure complete removal of the silver. In state 4 the micrometer is reversed and returns to the starting position at maximum speed. This state is terminated by the upper-limit switch (S2) mounted on the slide, which returns the device to state 0.

Controller operation

Sequential operation through the five states is accomplished by an electronic controller whose major subsystems are shown in Figure A3. The operating state of the device is determined by a sequential state controller consisting of the clock, state counter, state decoder, and state input multiplexer. The output of the state decoder determines the action performed by the motor speed control and directional control circuits. The state input

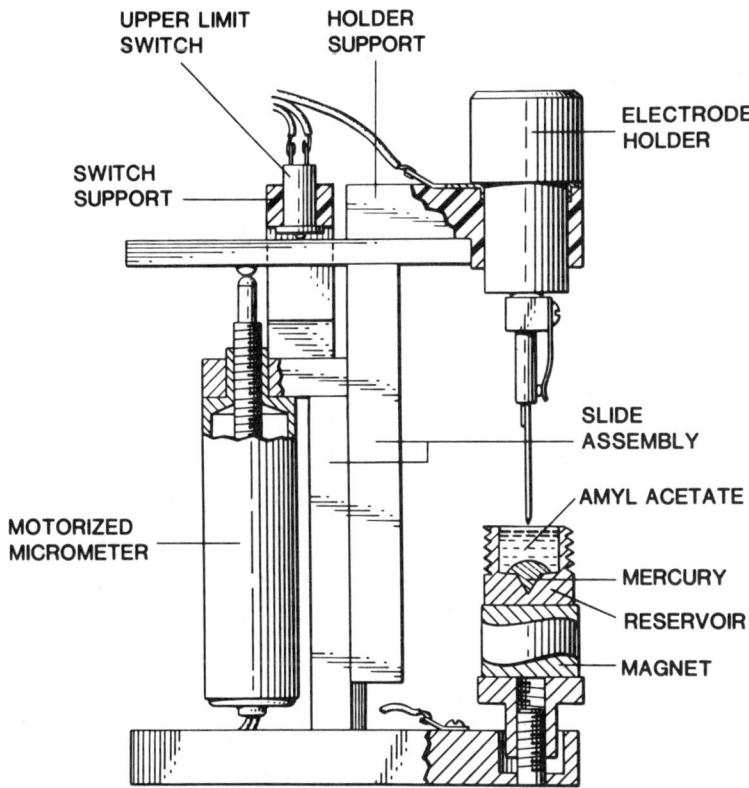

Fig. A1. Mechanical structure of electrode dipper. Motorized micrometer drives dovetail slide vertically. Electrode holder is made of brass and is supported by Lucite block for electrical insulation. Mercury is held in steel cup, which is threaded to attach dust cap. Reservoir is held to base by magnet so that it can be easily positioned and removed for cleaning. Mercury is covered with amyl acetate to remove organic material that might prevent intimate contact with mercury. Upper-limit switch detects end of return phase after silver removal.

multiplexer selects the input that terminates any given state. States of controlled duration (states 1 and 3) are terminated by an input from the interstate timing circuit. Other states are terminated by inputs from the switches or the electrode-contact sense circuit.

The motor speed control deserves special consideration. To obtain closed-loop speed control of a motor, a speed command signal is compared with a tachometer signal and the difference between these signals is applied to the motor after amplification. The tachometer signal may be obtained in various ways, typically requiring that an external sensing device be mounted on the motor shaft.

The physical construction of the motorized micrometer (see Fig. A1) precluded the use of any practical external tachometer. However, any electric motor produces a reverse electromotive force (EMF) proportional to motor speed. If the driving voltage is applied in pulses, this reverse EMF is available as a measure of motor speed during the interval between pulses. In the controller we compare the reverse EMF with a control signal and

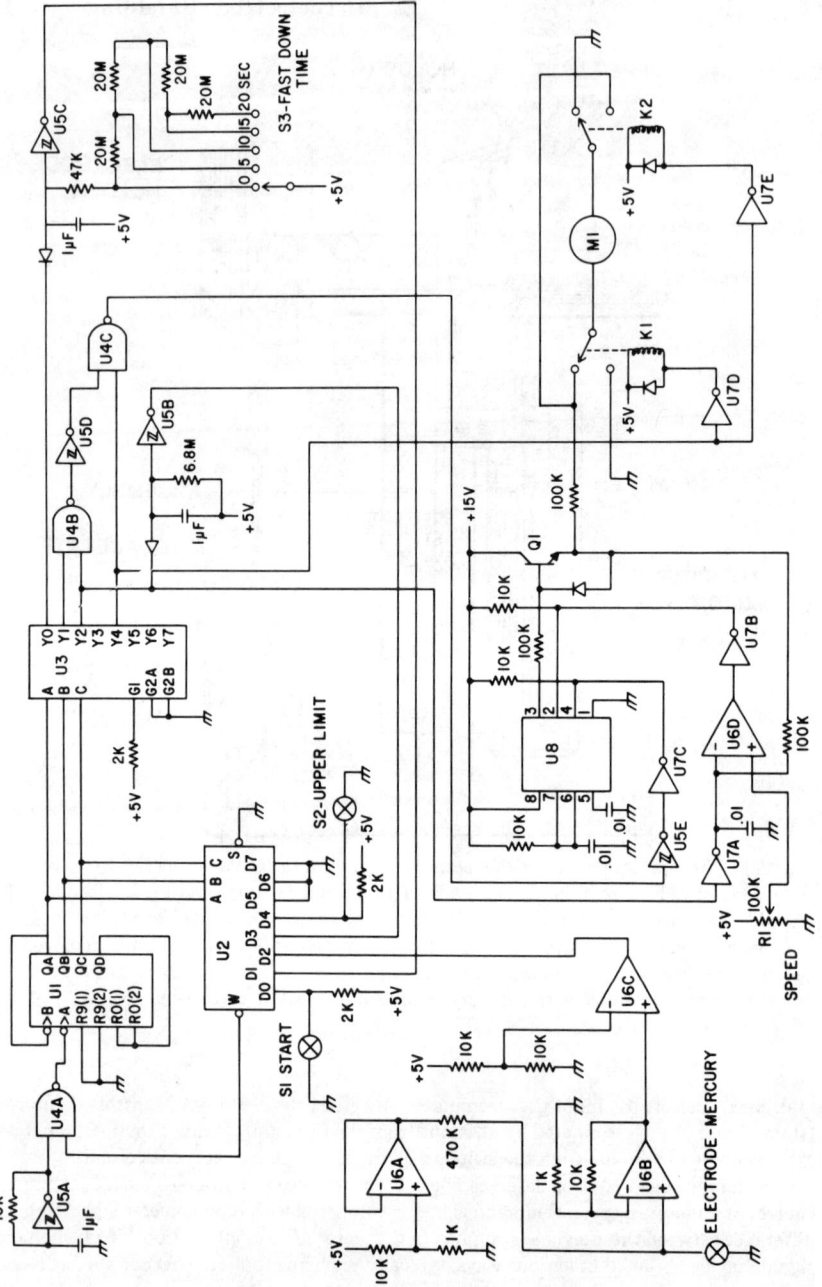

Fig. A2. Controller. Power supply is not shown—circuit requires +5 V for logic and op amps and +15 V for motor drive circuits. Active components: Q1, NPN silicon transistor (2N3054); U1, decade counter (74LS90); U2, multiplexer (74LS151); U3, decoder (74LS138); U4, quad nand gate (74LS00); U5, hex inverting Schmitt trigger (CD40106B); U6, quad single-supply op amp (LM324); U7, hex open collector inverter (7406); U8, monostable multivibrator (NE555).

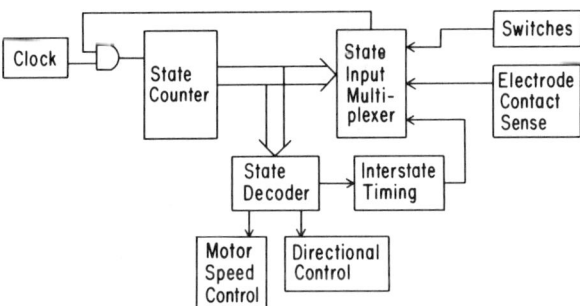

Fig. A3. Block diagram of controller. Summary of controller states. State 0: stopped, waiting for start switch. State 1: fast down, waiting for timeout. State 2: slow down, waiting for contact with mercury. State 3: stopped in mercury, waiting for timeout. State 4: fast up, waiting for top switch.

generate a driving pulse of fixed duration and amplitude when the reverse EMF falls below the control signal. Thus the output to the motor is pulse-position modulated, a crude form of frequency modulation.

Detailed circuit description

The state controller consists of the state counter (U1), a state decoder (U3), and a state input multiplexer (U2). Inputs to the multiplexer and outputs from the decoder are true for logic 0.

The start (S1) and upper-limit (S2) switches are contact closures to ground with pull-up resistors. Contact between the electrode and the mercury is conditioned by operational amplifiers U6a, U6b, U6c, and associated components.

Interstate timing (adjustable for state 1 and fixed for state 3) is provided by CMOS-inverting Schmitt triggers U5b and U5c, which are configured as inverting negative-pulse stretchers.

Micrometer speed is controlled by pulse-position modulation of U8, an NE555 monostable multivibrator. In the high-speed mode, U8 is continously retriggered, causing motor drive transistor Q1 to conduct continuously. In the slow-speed state, operational amplifier U6d compares the voltage at the emitter of Q1 with the reference voltage from potentiometer R1. When the reverse EMF produced by the motor falls below the reference voltage, the timing cycle of U8 is started by a negative pulse applied to the trigger input (pin 2). The motor is stopped by applying a logic-0 signal to the reset input of U8 (pin 4). Micrometer direction is controlled by relays K1 and K2, which are driven by open-collector inverters U7d and U7e.

Results

The device has been in experimental use and reliably produces electrodes shielded to within 10 μm of the tip. The standard deviation is not more than a few micrometers. The only problem that has arisen has been caused by oxidation of the silver coating on the electrode blanks. This oxidation raises contact resistance to the mercury and inhibits dissolution of the shield. The problem can be eliminated by storing the coated electrodes in a dessicator under vacuum. Electrodes stored in this way are stable for months.

FOUR

Conventional Voltage Clamping With Two Intracellular Microelectrodes

Alan S. Finkel
Axon Instruments, Inc., Burlingame, California

Peter W. Gage
Department of Physiology, The John Curtin School of Medical Research, Australian National University, Canberra, Australia

---

Terminology and Methods: Glossary of symbols, Potentials, currents, and null potentials, Resistances and conductances—time constants and frequencies, Laplace transforms, Transfer functions, block diagrams, and stability, Cell model • **Basic Description** • **Ideal Voltage Clamp: Membrane Capacitance as Sole Frequency-Dependent Component:** Membrane voltage responses to changes in command potential, Equivalent circuit, Membrane current responses, Conclusion • **Real Voltage Clamps:** Finite bandwidth of electronic amplifying circuit, Added phase lead to compensate amplifier input time constant, Effects of series resistance, Interelectrode capacitance, Changes in bath potential • **General Considerations:** Sensitivity to component variation, Slowing the rise time of command potential, Background noise under voltage clamp, Transfer-conductance voltage clamp • **Discussion**

When a conductance change occurs in an area of membrane that is small compared with the space constant of a cell, it is possible to control the membrane potential in that area using microelectrodes to monitor potential and to pass current. This technique was first used successfully by Takeuchi and Takeuchi (23) to record currents generated by acetylcholine at the motor end plate.

In this chapter the theory of electronic apparatus that may be used to voltage clamp a membrane area with two microelectrodes is discussed. Part of the theory of voltage clamping with two microelectrodes has been described by Katz and Schwartz (12) and by Smith et al. (20). Some of the principles described here have been well covered by these authors. However, for the sake of completeness, relevant equations are rederived with the Laplace transform technique. The notation and techniques we use are those of electrical circuit theory (6) and control theory (7).

The Laplace transform method is a frequency-domain technique well suited to the consideration of multiple variables. Initially some simple experimental arrangements are analyzed and previously reported results are confirmed. The analysis is then extended to consider the effects of stray capacitances, the series component of the cell membrane impedance, changes in impedance values, changing bath potentials, noise, amplifier frequency response, current measurement, and other topics. Where possible, recommendations for improved design and practical implementation of voltage-clamp circuits are made.

The performance of the voltage clamp and its dependence on the variables discussed in this chapter are illustrated with a simple cell model. This simple model was chosen because the complications of axons, dendrites, and electrical coupling would invalidate the generality of the analysis. Also, using a model rather than a real preparation allows variables such as membrane conductance and microelectrode resistance and capacitance to be altered at will. The question of any inadequacy of the spatial clamp because of the distributed nature of the cell membrane does not arise in this analysis because the model consists of simple electrical components. The problems associated with achieving an adequate space clamp are discussed in the chapters by Rall and by Kass and Bennett. The model chosen has characteristics typical of a tissue-cultured muscle cell.

Terminology and Methods

Glossary of symbols

Capacitance

C_{eff} effective coupling capacitance after neutralization
C_{in} capacitance at input of ME_1 buffer amplifier
C_m membrane parallel capacitance
C_μ feedback capacitor around clamp amplifier
C_n neutralization capacitor
C_{sh} shield capacitance
C_x coupling capacitance between microelectrodes
C_z zero-setting capacitor

Conductance

G conductance = $1/R$; subscripts have meaning assigned in definition of R
G_m^* $G_m + G_a$

G_T transfer conductance

General

A_v	voltage gain
B	feedback variable
$F(s)$	forward-path transfer function
f	cyclic frequency
f_{-3}	−3-dB frequency
H_f	gain of clamp feedback pathway
$H(s)$	feedback-path transfer function
H_{ss}	gain of R_s compensation circuit
I_m	membrane current
i_n	noise current source; subscripts have meanings assigned in definition of e_n
j	$\sqrt{-1}$
K	DC attenuation of V_o due to cell load
k	Boltzmann constant
ME	microelectrode
s	complex frequency variable
T	absolute temperature
t	time variable
t_r	10%–90% rise time
X	unspecified variable
Y	unspecified variable
Z	complex impedance; subscripts have meaning assigned in definition of R
α	low-frequency clamp gain
α_G	low-frequency gain of transfer-conductance voltage clamp
ϵ	error signal
ξ	damping ratio
μ	clamp amplifier gain
μ_s	gain of R_s compensation circuit
μ_x	cross-capacitance neutralization gain
σ	real part of s
ω	imaginary part of s; radial frequency
ω_{phys}	frequency of poles due to physical realization
1	subscript refers to voltage-recording microelectrode
2	subscript refers to current-passing microelectrode

Resistance

R	resistance
R_a	activatable membrane resistance
R_B	bulk resistance of bathing solution

R_e electrode resistance
R_{eq} source resistance of equivalent circuit
R_s series resistance = $R_{ss} + R_x$
R_{ss} series resistance due to preparation and bathing fluid
R_m membrane parallel resistance
R_μ gain-setting resistor
R_x coupling resistance between two microelectrodes
R_z zero-setting resistor

Time constant

τ time constant
τ_a decay time constant of activatable conductance
τ_B time constant of low-pass filter in bath-electrode circuit
τ_c command time constant
τ_{eq} equivalent time constant = $R_{eq}C_m$
τ_f time constant of R_s compensation circuit
τ_G predominant time constant of transfer-conductance voltage clamp
τ_I time constant of current-measurement circuit
τ_{IN} augmented input time constant = $R_{e1}(C_{in} + C_x)$
τ_{in} input time constant = $R_e C_{in}$
τ_m membrane time constant = $R_m C_m$
τ_p parallel time constant = $(R_m \parallel R_{e2})C_2$
τ_s series time constant = $(R_m \parallel R_s)C_m$
τ_μ pole time constant = $R_\mu C_\mu$
τ_x cross-capacitance time constant = $R_{e1}C_x$
τ_z zero time constant = $R_z C_z$
τ_0 predominant time constant
τ_0^* predominant time constant while G_a is nonzero

Voltage

E_a null (reversal) potential of activated ionic pathway
E_{RMP} resting membrane potential
e_n noise voltage source
$e_{n,eq}$ noise voltage source in equivalent circuit
$e_{n,m}$ noise voltage source due to membrane resistance
V_B potential of bathing solution
V_B' component of V_B due to extraneous factors
V_c command voltage
V_{eq} source voltage of equivalent circuit
V_m membrane potential, membrane potential difference
V_m^- initial condition of V_m
V_{mm} $V_m + V_s$

V_o output voltage of clamp amplifier
V_s voltage drop across R_s
V_{ss} feedback signal proportional to V_s
V_1 voltage recorded by ME_1

Potentials, currents, and null potentials

Throughout the text membrane potential (V_m) is used to refer to the potential difference across the membrane, inside with respect to outside, as is common practice. However, it is unlike some other potentials referred to in the text that are absolute values measured with respect to the ground potential of the electronic circuitry. The membrane current (I_m) is positive when positive ions flow from the inside to the outside of the cell. The null potential (reversal potential) is the value of V_m at which $I_m = 0$.

Resistances and conductances—time constants and frequencies

The letter R is always used to represent resistance and G is used to represent conductance; G and R are interchanged freely, with $R = G^{-1}$. Subscripts identify the particular application.

Time constants are always represented by τ. Generally, $\tau = RC$ (where C is capacitance); τ is related to radial frequency (ω) by $\tau = \omega^{-1}$ and to cyclic frequency (f) by $(2\pi f)^{-1}$.

Laplace transforms

Most of the analysis is done in the frequency domain with Laplace transforms (see ref. 7). The Laplace transform of a time-dependent variable $V(t)$ is

$$V(s) = \int_0^\infty V(t)e^{-st}dt$$

where t is the time variable and s is the complex frequency variable ($s = \sigma + j\omega$, where σ and ω are real variables and $j = \sqrt{-1}$).

Inverse Laplace transforms are found from tables. The Laplace transform of a variable is indicated by showing its functional dependence on s. Where no ambiguity is possible, the variable will frequently be written without showing its functional dependence on s or t.

Transfer functions, block diagrams, and stability

Transfer functions. Let $X(s)$ be an input to a linear time-invariant system; let $Y(s)$ be its output. The transfer function [TF(s)] of the system is

$$\text{TF}(s) = \frac{Y(s)}{X(s)}$$

Some of the important properties of the transfer function are listed next. *1)* The transfer function of a system is the Laplace transform of its impulse response. *2)* The roots of the denominator polynomial are the system poles; the roots of the numerator polynomial are the system zeros. *3)* The denominator is called the characteristic function of the system. The characteristic function alone determines the exponential and/or sinusoidal terms of the impulse response and the stability of the system.

Block diagrams. A block diagram is a graphical representation of a physical system. The block diagram of a negative-feedback system is shown in Figure 1. The rectangular blocks represent transfer functions, the circle represents a summing junction.

The closed-loop transfer function (CLTF) is derived directly from the block diagram

$$\text{CLTF}(s) = \frac{Y(s)}{X(s)} = \frac{F(s)}{1 + F(s)H(s)} \qquad (1)$$

where $F(s)$ is the forward-path transfer function and $H(s)$ is the feedback-path transfer function. The open-loop transfer function (OLTF) is the transfer function around the loop when the loop is broken at any one point.

$$\text{OLTF}(s) = F(s)H(s)$$

Stability. Each zero or pole in the OLTF contributes up to $+90°$ or $-90°$, respectively, to the open-loop phase shift and up to $+20$ dB/decade or -20 dB/decade, respectively, to the attenuation of the open-loop gain. For any negative-feedback circuit, stability (in the sense that circuit does not oscillate) is assured if the excess phase shift in the open-loop circuit is $<180°$ at the frequency at which $F(s)H(s)$ crosses through unity gain (Nyquist's criterion restated; see ref. 4, p. 501). An excess phase shift of $180°$ corresponds to positive feedback.

Because of the interdependence of the gain and phase characteristics (3), a total phase shift approaching $180°$ is associated with a rate of attenuation in the open-loop gain approaching 40 dB/decade. Conversely,

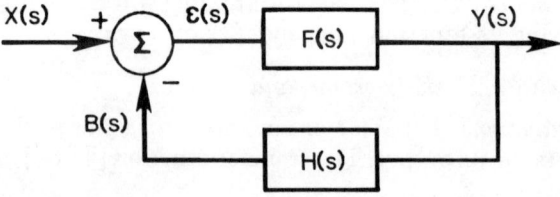

Fig. 1. Negative-feedback system. $F(s)$, forward-path transfer function; $X(s)$, input function; $Y(s)$, output function; $B(s)$, feedback function; $\epsilon(s)$, error function; $H(s)$, feedback-path transfer function.

and in conjunction with the stability criterion, if an asymptotic approximation to the gain response is falling at 40 dB/decade (or more) at the frequency at which the OLTF crosses through unity gain, then the closed-loop circuit will be liable to oscillate.

A Bode gain plot is a graph of the asymptotic log gain versus the log frequency of the OLTF. It is constructed by summing the magnitude contribution of each pole and zero. It allows a quick determination of how much DC gain can be used with a given placement of poles and zeros without causing instability.

Cell model

The cell model used in the analysis and testing of the voltage-clamp circuits is shown in Figure 2. The resting membrane of the cell consists of a resistance (R_m) and a capacitance (C_m) in parallel. The membrane time constant is $\tau_m = R_m C_m$. A battery (E_{RMP}) in series with R_m represents resting membrane potential. In parallel with these components is a variable resistance R_a, which is the sum of all activatable membrane resistances. The resting value of R_a is infinite because all the resting membrane resistance is attributed to R_m. On activation, R_a becomes finite and carries an ionic current driven by the battery E_a, which represents the null potential for the activated ionic pathways. The value of R_a is a

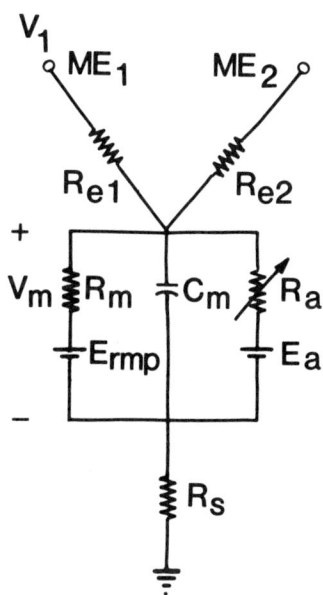

Fig. 2. Cell model used for voltage-clamp analysis and testing. R_m and C_m, cell resting membrane resistance and capacitance, respectively; E_{RMP}, battery representing resting membrane potential; R_a, activatable membrane resistance, which is infinite at rest but which becomes finite on activation and carries ionic current driven by battery E_a; R_s, resistance in series with membrane; ME_1 and ME_2, voltage-recording and current-passing microelectrodes of resistance R_{e1} and R_{e2}, respectively; V_1, potential recorded by ME_1; V_m, membrane potential.

function of voltage and time; R_s is a resistance in series with the membrane.

The voltage-recording microelectrode (ME_1) and the current-passing microelectrode (ME_2) are represented by resistors R_{e1} and R_{e2}, respectively. The potential (V_1) recorded by ME_1 is not necessarily equal to V_m. If there is no voltage drop across R_s (either because $R_s = 0$ or $I_m = 0$) and R_{e1} (because ME_1 is connected to ideal buffer amplifer with zero bias current and input capacitance), then $V_1 = V_m$. This equivalence is assumed throughout the next two sections.

Without limiting generality, E_{RMP} is assumed to be equal to zero in the following analyses. Nonzero values of E_{RMP} may be incorporated by adding the value of E_{RMP} to the calculated value of V_m. The values used to represent the membrane are $R_m = 10$ MΩ and $C_m = 1$ nF. These are representative of a tissue-cultured muscle cell (20). Standard values used for the microelectrodes are $R_{e1} = R_{e2} = 10$ MΩ. When the specific details of the membrane are not essential to the theory, the cell membrane is represented by its complex membrane impedance (Z_m).

Basic Description

The aim of a voltage-clamp circuit is to control the membrane potential of a cell. At the same time the current required for this control is measured. Because both membrane potential and current are known, the membrane impedance can be directly calculated.

Figure 3 shows a simplified voltage clamp with two microelectrodes. The value of V_m measured by an intracellular microelectrode (ME_1) is summed with a command voltage (V_c) in a high-gain DC-coupled differential amplifier. The voltage gain (μ) of the amplifier is typically $>10^3$. The output voltage (V_o) causes current (I_m) to flow via a second microelectrode (ME_2) into the cell and out across Z_m, thereby changing V_m (feedback variable). The output voltage continuously adjusts to keep $V_m = V_c - V_o/\mu$. For large values of μ, V_m effectively equals V_c. Current I_m is measured by a current-measurement circuit, shown in the figure as a meter and discussed in detail in the chapter by Finkel. In practice, additional features must be designed into the circuit for it to work satisfactorily.

There are several essential features of a good voltage clamp. *1)* Fidelity: the gain (μ) must be large so that V_m (measured at tip of ME_1) closely follows V_c even when the membrane conductance or the resistance of ME_2 undergoes a large change. *2)* Dynamic response: the clamp must establish new potentials quickly. The current required to charge the membrane capacitance should approach zero before the onset of any membrane conductance changes. Equivalently the clamp must be able to follow rapid changes in membrane conductance at fixed membrane

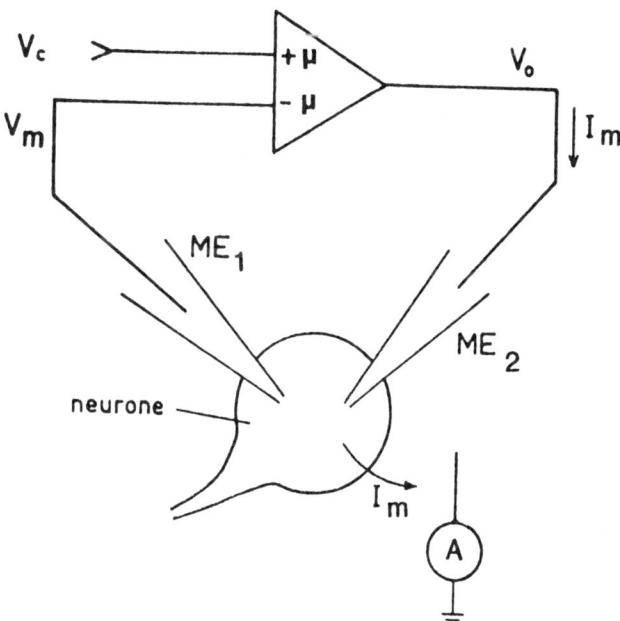

Fig. 3. Simplified two-electrode voltage clamp. High-gain (μ) differential amplifier acts to keep differential voltage at its input (= $V_c - V_m$) small by passing current into or out of cell. V_c, command potential; V_o, output of differential amplifier; V_m, membrane potential; I_m, current through cell membrane, measured by ammeter (circular device) that ideally has zero resistance and therefore no effect on circuit.

potentials. 3) Accuracy: the clamp should control V_m alone. Any potentials occurring in the tissue fluid and potentials caused by an ohmic voltage drop across any series resistance between the tip of ME_1 and the grounding point of the preparation must be rejected.

The dynamic response of a feedback circuit is determined by the number and position of the frequency-dependent circuit components. The frequency-dependent components, which will be discussed in the next sections, are 1) C_{in}, the lumped equivalent of the capacitance to ground at the input of the ME_1 buffer amplifier and through the wall of ME_1; 2) C_z, a variable capacitor that can be introduced into the circuit to provide phase lead (i.e., a zero); 3) C_μ, a variable capacitor that can be introduced into the circuit to provide phase lag (i.e., a pole); 4) C_x, the coupling capacitance between the microelectrodes; and 5) C_m, the parallel capacitance of the cell membrane.

The simplest voltage-clamp configuration for analysis arises when there is only one frequency-dependent component that, because of the application of the circuit, must be C_m. This configuration is examined

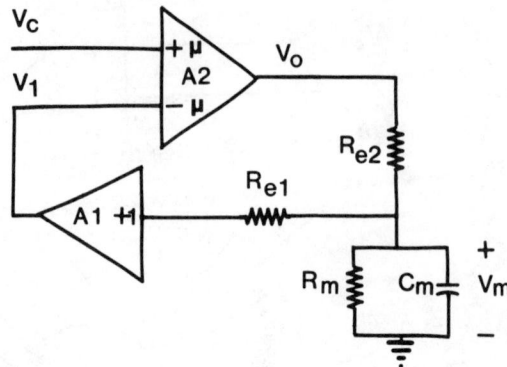

Fig. 4. Schematic circuit of two-electrode voltage clamp. Unity-gain feedback path is R_{e1} and A1 in series. Forward path consists of differential amplifier (A2) and attenuator formed by R_{e2} in series with cell. V_o, output of A2; V_m, output of forward path; V_1, feedback variable.

next. The other frequency-dependent components may be sufficiently removed or compensated for so that they have insignificant effect over a limited range of clamp gains. The clamp performance after these frequency-dependent components are reintroduced is considered subsequently.

Ideal Voltage Clamp: Membrane Capacitance As Sole Frequency-Dependent Component

In its simplest configuration (Fig. 4) the voltage-clamp circuit consists of a cell at rest, a buffer amplifier with high input impedance (A1), and a variable-gain DC-coupled amplifier (A2).

Membrane voltage responses to changes in command potential

The feedback path in Figure 4 is simply R_{e1} and A1 in series. Because A1 is ideal (i.e., infinite input impedance), the feedback is unity gain so that $V_1 = V_m$ and $H(s) = 1$.

The forward path $[F(s)]$ consists of A2, R_{e2}, and the cell load. The input function to $F(s)$ is the error voltage, $V_c(s) - V_m(s)$. The output function is $V_m(s)$. Thus

$$F(s) = \frac{V_m(s)}{V_c(s) - V_m(s)}$$

$$= \frac{\mu K}{s\tau_p + 1} \qquad (2)$$

where

$$K = \frac{R_m}{R_m + R_{e2}}$$

$$\tau_p = (R_m \parallel R_{e2})C_m$$

the symbol ∥ means in parallel with, K is the DC attenuation of V_o (output of A2) caused by the cell load and R_{e2}, and τ_p is the time constant for charging the cell through R_{e2}.

Substituting $F(s)$ and $H(s)$ into Equation 1 yields the CLTF

$$\frac{V_m(s)}{V_c(s)} = \frac{\mu K}{s\tau_p + \mu K + 1} \tag{3}$$

Fidelity. From the final-value theorem (5), the steady-state response of the circuit to a step change in V_c is found by letting s approach zero. That is

$$\lim_{s \to 0} \frac{V_m(s)}{V_c(s)} = \frac{\mu K}{\mu K + 1} = \alpha \tag{4}$$

where α is a measure of the fidelity with which the clamp circuit forces the membrane potential to follow the command voltage at DC and low frequencies. For very large values of μ, α approaches unity.

For later use it is convenient to reexpress the transfer function given in Equation 3 in standard form

$$\frac{V_m(s)}{V_c(s)} = \frac{\alpha}{s\tau_0 + 1} \tag{5}$$

where

$$\tau_0 = \frac{\tau_p}{\mu K + 1}$$

$$= \frac{R_m R_{e2} C_m}{(\mu + 1) R_m + R_{e2}}$$

and τ_0 is the predominant time constant describing the system impulse response.

Step response. The time-domain response to a step command (step response) at $t = 0$ is found by taking the inverse Laplace transform of Equation 5 after letting $V_c(s) = V_c/s$. Therefore

$$V_m(t) = \alpha V_c[1 - \exp(-t/\tau_0)] \quad t > 0 \tag{6}$$

Thus $V_m(t)$ rises exponentially with time constant τ_0 to the final value αV_c. The 10%–90% rise time is

$$t_r = 2.2\tau_0 \tag{7}$$

For $\mu K \gg 1$

$$t_r = 2.2 R_{e2} C_m / \mu \tag{8}$$

Thus, in this ideal voltage clamp, increasing the clamp gain decreases the rise time. Interestingly, as long as $\mu K \gg 1$, the cell membrane resistance is not a factor affecting t_r (Eq. 8).

In the cell model of Figure 2, $K = 0.5$ and $\tau_p = 5$ ms. If $\mu = 200$, the predominant time constant is $\tau_0 = 49.5$ s and the 10%–90% rise time is $t_r = 109$ μs.

Frequency response. Equation 5 is the same as the transfer function of a first-order low-pass filter with a −3-dB frequency (f_{-3}) at

$$f_{-3} = (2\pi\tau_0)^{-1} \qquad (9)$$

which, for $\mu K \gg 1$, is approximately

$$f_{-3} = \mu/2\pi R_{e2}C_m \qquad (10)$$

Thus, increasing the clamp gain directly increases the bandwidth with which V_m can follow changes in V_c.

Stability. Figure 5 (lower curve) shows the Bode gain plot for this example. The attenuation from the open-loop pole at τ_p^{-1} is 20 dB/decade for all frequencies $>\tau_p^{-1}$, including the frequency at which the gain crosses through unity (0 dB). The effect of increasing μK is to shift the gain curve upward without changing its shape. Thus the ideal circuit is unconditionally stable.

In practice, at very high values of μK, the rate of attenuation equals or exceeds 40 dB/decade whereas the open-loop gain remains greater

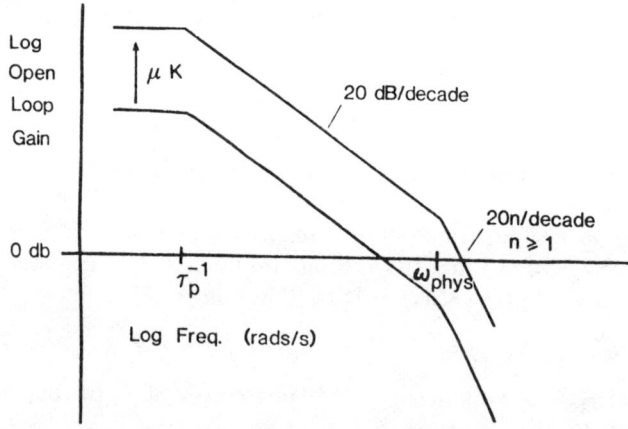

Fig. 5. Asymptotic Bode gain plot. Single open-loop pole at τ_p^{-1} (parallel time constant) causes attenuation at 20 dB/decade. In ideal case, multiple poles at ω_{phys} caused by physical realization of circuit are nonexistent. Thus attenuation of open-loop gain remains at 20 dB/decade and circuit is unconditionally stable. In practice, at high values of μK, n poles at ω_{phys} cause attenuation at unity-gain crossover frequency to be $20n$ dB/decade. This is unstable rate of attenuation. Therefore maximum usable gain is limited by unmodeled poles.

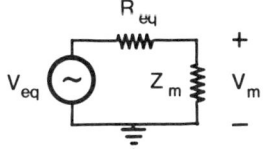

Fig. 6. Equivalent circuit of ideal voltage clamp consists of voltage source (V_{eq}) and resistor (R_{eq}) in series. Cell membrane impedance (Z_m) is load.

than unity. This is because unidentified high-frequency poles from the physical realization of the circuit become effective and destabilize the circuit. They are represented in Figure 5 as multiple poles at ω_{phys}, although in practice they are not necessarily at the same frequency.

Equivalent Circuit

The transfer function in Equations 3 and 5 can be reexpressed in terms of the membrane impedance (Z_m) to yield

$$\frac{V_m(s)}{V_c(s)} = \frac{\mu Z_m}{R_{e2} + (\mu + 1)Z_m} \quad (11)$$

The equivalent circuit shown in Figure 6 (known also as Thevenin equivalent circuit) consists of a voltage source (V_{eq}) and a source resistance (R_{eq}) in series with the load; V_{eq} is the value of V_m that would be measured if the cell load was removed

$$V_{eq}(s) = \lim_{Z_m \to \infty} V_m(s) = \frac{\mu}{\mu + 1} V_c(s) \quad (12)$$

and R_{eq} is that value of the load impedance for which the output voltage drops to half of the open-circuit value

$$R_{eq} = Z_m(V_m = 0.5 V_{eq}) = \frac{R_{e2}}{\mu + 1} \quad (13)$$

Besides being a simplified representation of the voltage-clamp circuit as a voltage source in series with a resistor, the equivalent circuit is also a useful representation of the circuit for the solution of specific problems such as the current responses discussed in the next section. It can also be used to derive all of the performance equations that were previously derived directly. For example, from the equivalent circuit we get

$$\frac{V_m(s)}{V_{eq}(s)} = \frac{Z_m}{R_{eq} + Z_m} \quad (14)$$

Substituting Equations 12 and 13 yields

$$\frac{V_m(s)}{V_c(s)} = \frac{\alpha}{s\tau_0 + 1}$$

where

$$\alpha = \frac{\mu}{\mu + 1} \frac{R_m}{R_m + R_{eq}} = \frac{\mu K}{\mu K + 1}$$

and

$$\tau_0 = (R_m \| R_{eq})C_m = \tau_p/(\mu K + 1)$$

as given in Equations 4 and 5. For $\mu \gg 1$, such that $R_{eq} \to 0$

$$\tau_0 = R_{eq}C_m \tag{15}$$

When $\mu = 200$, $R_{eq} = 49.8$ kΩ for the cell model, while $\tau_0 \approx 49.8$ μs as before.

Membrane current responses

In a typical experimental situation, the step response of $V_m(t)$ is monitored while the voltage clamp is optimally adjusted. The experimenter increases the gain until the desired step response of $V_m(t)$ is achieved and then proceeds to measure $I_m(t)$.

In this section the dependence of I_m on each of several variables (X) is derived, and the question of how the measurement of I_m relates to the changes in X is answered. The relationship between the transfer function $V_m(s)/V_c(s)$, which is used by the experimenter to set up the voltage clamp, and the experimentally significant transfer function $I_m(s)/X(s)$ is shown. In each case $I_m(s)/X(s)$ is found using the equivalent-circuit representation connected to the cell model as the load (Fig. 7).

A general expression for the membrane current can be written directly from the circuit in Figure 7 by summing the currents through each parallel branch of the membrane and eliminating $V_m(s)$. That is

$$I_m(s) = \frac{V_{eq}(s)sC_m + V_{eq}(s)G_m + [V_{eq}(s) - E_a(s)]G_a(s)}{sC_mR_{eq} + G_mR_{eq} + G_a(s)R_{eq} + 1} \tag{16}$$

The three numerator terms decide the currents through C_m, G_m, and G_a. Equation 16 is the basis for the solution of the transfer function $I_m(s)/X(s)$ in the next sections.

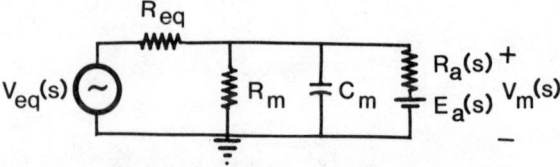

Fig. 7. Equivalent circuit used for deriving membrane current responses.

Membrane current caused by changes in command potential. For the membrane at rest ($G_a = 0$), Equation 16 simplifies to

$$\frac{I_m(s)}{V_{eq}(s)} = \frac{sC_mR_m + 1}{sC_mR_mR_{eq} + R_m + R_{eq}}$$

Substituting for $V_{eq}(s)$ from Equation 12 yields the desired transfer function

$$\frac{I_m(s)}{V_c(s)} = \frac{\alpha(s\tau_m + 1)}{R_m(s\tau_0 + 1)} \tag{17}$$

Fidelity. Applying the final-value theorem to Equation 17 yields the steady-state current

$$I_m(\infty) = \alpha V_c(\infty)/R_m \tag{18}$$

Step response. The response to a step command at $t = 0$ is found by letting $V_c(s) = V_c/s$ in Equation 17 and looking up the inverse Laplace transform in standard tables. This yields

$$I_m(t) = \frac{\alpha V_c}{R_m}\left[1 + \frac{\tau_m - \tau_0}{\tau_0}\exp(-t/\tau_0)\right] \quad t > 0 \tag{19}$$

After a step change in $V_c(t)$, $I_m(t)$ is an exponential that decays from an initial peak at $t = 0$ toward a steady-state value with the same time constant as the exponential rise of $V_m(t)$ (Eq. 6). For $\mu K \gg 1$, $\tau_0 \ll \tau_m$; thus the initial peak at $\alpha V_c\tau_m/R_m\tau_0$ is much greater than the steady-state value. The relationship between $I_m(t)$ and $V_m(t)$ at different values of μ and R_m is illustrated in Figure 8A.

Measurement of membrane impedance. The most frequent reason for measuring the transfer function $I_m(s)/V_c(s)$ is to derive the complex membrane impedance $[Z_m(s)]$. From Figure 6 it can be seen that

$$Z_m(s) = \frac{V_{eq}(s)}{I_m(s)} - R_{eq}$$

Normally the clamp would be operated at high gain so that $Z_m(s) \approx V_c(s)/I_m(s)$. The waveforms commonly used for $V_c(s)$ during impedance measurements are sinusoidal, triangular, and square waveforms (17, 24). The reasons for using any particular waveform depend on the expected characteristics of Z_m and are discussed in detail by Palti and Adelman (17).

Membrane current caused by activatable membrane conductance. In the previous section it was shown how the voltage-clamp technique can be used to measure the complex membrane impedance. However, the voltage-clamp technique is more often used to hold V_m constant and to

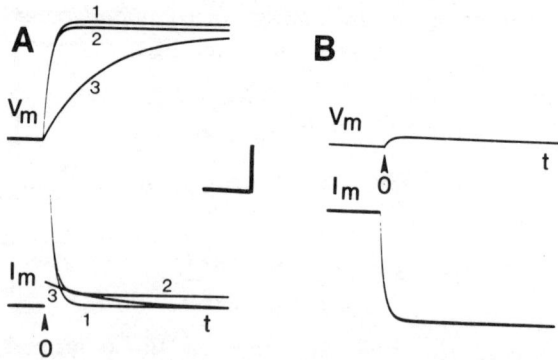

Fig. 8. Comparison of step response of various transfer functions for ideal voltage clamp. A: *upper traces,* $V_m(t)$ after 100-mV step in $V_c(t)$ at $t = 0$; *lower traces,* $I_m(t)$ recorded simultaneously; C_m was 1 nF and $R_e = R_m = 10$ MΩ in all cases. Parameters varied were *1)* $\mu = 200$, $R_a = \infty$; *2)* $\mu = 200$, $R_a = 1$ MΩ; and *3)* $\mu = 20$, $R_a = \infty$. Decay time constants of $V_m(t)$ and $I_m(t)$ are same. Responses are described by Eqs. 6 and 19. *B:* response to step change in $R_a(t)$. V_c was held constant. $C_m = 1$ nF and $R_e = R_m = 10$ MΩ; $\mu = 200$ and $V_c = -100$ mV. At $t = 0$, R_a was stepped from ∞ to 1 MΩ. Time constant for establishing new current was same as for *trace 2* in A. Shift in V_m is caused by decrease in α resulting from greater attenuation caused by cell load (Eqs. 2 and 4).

measure I_m as G_a or E_a change. If V_m is held constant, the cell membrane capacitance makes no contribution to the measured current. The changes in G_a may arise from voltage activation caused by a previous step change in V_c (e.g., voltage-jump relaxation analysis) or from chemical activation occurring while V_c is held constant (e.g., fluctuation analysis and the measurement of synaptic currents).

Step response. When $G_a(s)$ is nonzero, Equation 16 cannot be simplified before the inverse Laplace transform is taken. Nor can the transfer function be written in the form $I_m(s)/G_a(s)$ because $G_a(s)$ appears in the denominator polynomial and in the numerator polynomial. Instead $I_m(s)$ must be written as a function of all variables and solved for the particular input function.

If $G_a(t)$ steps at $t = 0$ from zero to the value G_a, the circuit can be solved by letting $G_a(s) = G_a$ and including the initial condition. The initial condition is the steady-state value of $V_m(t)$ immediately prior to the step. This value (V_m^-) must be incorporated into the expression for $I_m(s)$ by putting a voltage source V_m^-/s in series with C_m. The other voltage sources are nonzero at $t = 0$ and are represented by $V_{eq}(s) = V_{eq}/s$ and $E_a(s) = E_a/s$. Therefore Equation 16 becomes

$$I_m(s) = \frac{(V_{eq} - V_m^-)C_m}{sC_mR_{eq} + G_m^*R_{eq} + 1} + \frac{V_{eq}G_m + (V_{eq} - E_a)G_a}{s(sC_mR_{eq} + G_m^*R_{eq} + 1)} \quad (20)$$

where $G_m^* = G_m + G_a$ and $V_m^- = R_mV_{eq}/(R_{eq} + R_m)$. From standard tables the inverse Laplace transform is found to be

Two-Microelectrode Voltage Clamping

$$I_m(t) = \frac{I_1 \tau_0^*}{\tau_{eq}} + (I_2 - I_1 \tau_0^*/\tau_{eq}) \exp(-t/\tau_0^*) \quad t > 0 \quad (21)$$

where

$$I_1 = V_{eq} G_m + (V_{eq} - E_a) G_a$$

$$I_2 = (V_{eq} - V_m^-)/R_{eq}$$

$$\tau_{eq} = R_{eq} C_m$$

and

$$\tau_0^* = (R_{eq} \parallel R_m^*) C_m$$

Comparing Equation 21 with Equation 6 it can be seen that the time constant τ_0^* governing the settling of $I_m(t)$ after a step change in $R_a(t)$ is the same as the time constant governing settling of $V_m(t)$ after a step change in V_c if R_m had a value $R_m^* = R_m \parallel R_a$. This is illustrated in Figure 8B, which shows $I_m(t)$ resulting from a step change in R_a to the same value used in Figure 8A.

If $G_a(t)$ was constant and $E_a(t)$ was stepped at $t = 0$, $I_m(t)$ would have the same form as in Equation 21. The only difference would be in the value of V_m^-.

The solution of the various transfer functions $[I_m(s)/X(s)]$ is much simpler if the open-loop gain is large. This is shown in the next section.

Membrane current responses at large gains. At large clamp gains, $R_{eq} \cdot [G_m + G_a(s)] \ll 1$ so that Equation 16 becomes

$$I_m(s) = \frac{V_{eq}(s) s C_m + V_{eq}(s) G_m + [V_{eq}(s) - E_a(s)] G_a(s)}{s R_{eq} C_m + 1} \quad (22)$$

Changes in activatable membrane conductance. When $V_{eq}(t)$ is a constant, the contribution to $I_m(t)$ of the first term in Equation 22 is zero and the contribution of the second term is a constant equal to the steady-state value

$$I_2 = V_{eq} G_m \quad (23)$$

If $E_a(s)$ is also constant, the third term yields

$$\frac{I_3(s)}{G_a(s)} = \frac{V_{eq} - E_a}{s R_{eq} C_m + 1} = \frac{V_{eq} - E_a}{s \tau_0 + 1} \quad (24)$$

This equation has a constant numerator and exactly the same characteristic function as Equation 5, which describes the transfer function $V_m(s)/V_c(s)$. Thus the bandwidth and step response for measuring changes in $I_m(t)$ as a function of $G_a(t)$ are exactly the same as the bandwidth and step response for measuring changes in $V_m(t)$ as a function of $V_c(t)$.

In some experiments (e.g., fluctuation analysis) it is important to set the clamp bandwidth (as described by Eq. 24) to be greater than the bandwidth of the conductance fluctuations. Equivalently, in experiments such as synaptic current analysis, the step response of $I_3(t)$ must be made brief compared with the synaptic conductance transient. For example, if the time function of synaptic conductance change is described by

$$G_a(t) = G_a \exp(-t/\tau_a) \quad t \geq 0$$
$$= 0 \quad\quad\quad\quad\quad t < 0$$

then from standard tables of Laplace transforms

$$G_a(s) = G_a(s + 1/\tau_a)^{-1}$$

Substituting $G_a(s)$ into Equation 24 and taking inverse Laplace transforms yields

$$I_3(t) = \frac{G_a(V_{eq} - E_a)}{1 - \tau_0/\tau_a} [\exp(-t/\tau_a) - \exp(-t/\tau_0)]$$

If $\tau_a \gg \tau_0$, the rise time is governed by τ_0 whereas the decay is dominated by τ_a. If $\tau_a < \tau_0$, the decay and the rise will be dominated by τ_0 and the estimate of $G_a(t)$ will be grossly erroneous.

Changes in driving force. If $G_a(t)$ and $V_{eq}(t)$ are constants but $E_a(t)$ is not, the transfer function $I_3(s)/E_a(s)$ also has the same form as the transfer function $V_m(s)/V_c(s)$.

Conclusion

The transfer functions describing the important input-output relationships of the ideal voltage-clamp circuit have been developed and analyzed. The easiest transfer function to set up experimentally and to calculate theoretically is $V_m(s)/V_c(s)$. This transfer function has the same characteristics as a first-order low-pass filter with input V_c and output V_m. The low-frequency gain is always less than unity and depends on the magnitude of the clamp gain and the attenuation caused by the voltage divider formed by the resistances of ME_2 and the cell (Eq. 4). The high-frequency behavior is governed by the predominant time constant (τ_0).

The stability of the ideal voltage clamp is guaranteed even at infinitely high clamp gain. This happy situation is a result of the assumptions that the electronic and microelectrode responses are infinitely fast.

An equivalent circuit has been derived (see Fig. 6). This circuit is a concise representation of the voltage-clamp circuit, enabling easier visualization of its operation. In addition the equivalent circuit makes it possible to calculate transfer functions in a straightforward manner.

For all the transfer functions the step responses consist of the same exponential functions with different coefficients and constant terms. The step response of $I_m(s)/G_a(s)$ is identical in shape to the step response of $V_m(s)/V_c(s)$, provided the latter is derived at the new value of $G_a(t)$. Thus, if the voltage-clamp circuit is set to give a certain fidelity and dynamic response for $V_m(t)$ as a function of $V_c(t)$, the same fidelity and dynamic response is achieved for the measurement of $I_m(t)$ as a function of $G_a(t)$ or $E_a(t)$.

This equivalence is generally true and applies to the more complicated voltage clamps discussed in the next sections, which have more than one frequency-dependent component. The only restriction is that G_a must be in parallel with the whole of the preparation across which V_m is measured. The reason for the equivalence is that the different transfer functions have been calculated by rearranging the input-output equations without changing the actual system. These rearrangements do not change the denominators of the transfer functions. The denominator alone determines the exponential components of the time-domain step response (7). The numerator determines the coefficients and the phase of these components but does not affect the basic nature of the response.

This equivalence is extremely convenient because it means that in all cases the experimenter can set the voltage-clamp parameters to give an optimum step response in $V_m(t)$ to step changes in $V_c(t)$ and know that the fidelity and decay time constant observed during this set-up procedure are exactly the same as the fidelity and decay time constant of the experimental phenomena under study. The only restriction on the generality of equivalence is that the clamp gain must be sufficiently high, or changes in G_a must be sufficiently small, for changes in G_a not to affect the denominator.

In the cases described next, only the transfer function $V_m(s)/V_c(s)$ will be calculated.

Real Voltage Clamps

In a real voltage clamp many factors arise to affect adversely the performance described in the previous section. These include the presence of a resistance (R_s) in series with the cell membrane, undesirable shifts in the bath potential, microelectrode input capacitance, finite bandwidth of the electronics, and coupling resistance and capacitance between the microelectrodes. The aim of this section is twofold. *1)* Techniques to compensate for the adverse factors are described. In theory these compensations cause the voltage clamp to revert to the behavior of the ideal clamp. *2)* The behavior of the voltage clamp in the presence of the adverse factors is described. The adverse factors are treated individually.

Some of the factors discussed in this section are contained in the circuit shown in Figure 9A. Besides C_m, there are three other frequency-dependent components. The stray capacitance at the input of the voltage-

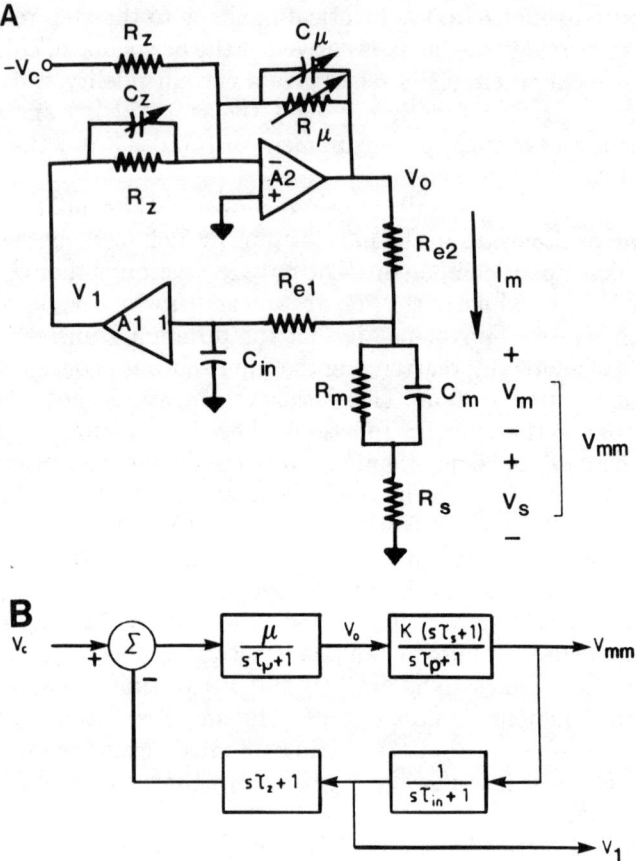

Fig. 9. *A*: wide-band voltage-clamp circuit. *B*: transfer-function diagram of wide-band voltage-clamp circuit.

recording amplifier and the distributed leakage capacitance through the wall of the microelectrode to the bathing solution have been lumped together and represented by the single capacitor C_{in}. The input time constant is $\tau_{in} = R_{e1}C_{in}$; C_μ is a capacitor limiting the high-frequency bandwidth of the summing amplifier (A2). The A2 time constant is $\tau_\mu = R_\mu C_\mu$. Even if C_μ is not deliberately built into the circuit, it is still present in the analysis to represent the finite bandwidth of the electronics. A capacitor (C_z) may be added to the circuit to boost the high-frequency summation of V_m. This is known as placing a "zero" or a "phase lead" into the circuit. The time constant of this phase-lead circuit is $\tau_z = R_z C_z$.

An additional component in this circuit is R_s, the series component of the membrane resistance. It affects the frequency response of the membrane by limiting the maximum high-frequency current through the membrane. The time constant governing this effect is $\tau_s = (R_m \| R_s)C_m$. The voltage-recording microelectrode cannot distinguish between V_m and V_s (voltage across R_s) and therefore the clamp acts on their sum V_{mm}. It should be noted that, because the command potential is summed at the inverting input of the operational amplifier, negative commands produce positive changes in V_m. The gain of the summing amplifier is $\mu = -R_\mu/R_z$.

The transfer-function representation of the wide-band voltage-clamp circuit is shown in Figure 9B. For clarity, the forward-path gain is in two parts; the gain from the amplifier and the attenuation from the series combination of the current-passing microelectrode and the cell impedance. The transfer function is

$$\frac{V_{mm}(s)}{V_c(s)} = \frac{\mu K(s\tau_s + 1)(s\tau_{in} + 1)}{(s\tau_\mu + 1)(s\tau_p + 1)(s\tau_{in} + 1) + \mu K(s\tau_z + 1)(s\tau_s + 1)} \quad (25)$$

where

$$K = \frac{R_m + R_s}{R_m + R_s + R_{e2}}$$

and

$$\tau_p = [R_m \| (R_s + R_{e2})]C_m$$

The next discussions are based on this transfer function. It should be noted that for τ_s, τ_{in}, τ_μ, and $\tau_z = 0$ the transfer function is identical to the ideal transfer function in Equation 3. The same would also be true if pole-zero cancellation were used by setting $\tau_z = \tau_{in}$ and $\tau_\mu = \tau_s$.

Finite bandwidth of electronic amplifying circuit

In this section it is assumed that τ_z is set to cancel the pole caused by C_{in} (i.e., $\tau_z = \tau_{in}$) and that $\tau_s = 0$ so that $V_s = 0$ and $V_{mm} = V_m$. In this case the only additional term is from τ_μ and the transfer function simplifies to

$$\frac{V_m(s)}{V_c(s)} = \frac{\alpha}{s^2\tau_\mu\tau_p/(\mu K + 1) + s(\tau_\mu + \tau_p)/(\mu K + 1) + 1} \quad (26)$$

where $\alpha = \mu K/(\mu K + 1)$ as before. The inverse Laplace transform can be found from standard tables. Alternatively, tabulated data on the step response of all-pole second-order systems can be used if Equation 26 is expressed in the standard form (4, 5)

$$\frac{V_m(s)}{V_c(s)} = \frac{\alpha}{s^2\tau_0^2 + 2s\xi\tau_0 + 1} \quad (27)$$

where τ_0 is the predominant time constant and ξ is the damping ratio (measure of signal transient performance). Equating coefficients in Equations 26 and 27 yields

$$\tau_0 = \sqrt{\frac{\tau_\mu \tau_p}{\mu K + 1}}$$

and

$$\xi = \sqrt{\frac{(\tau_\mu + \tau_p)^2}{4\tau_\mu \tau_p(\mu K + 1)}} \tag{28}$$

If $\xi > 1$, the response is overdamped. If $0 < \xi < 1$, the response is underdamped (i.e., step response overshoots). At $\xi = 1$, the response is critically damped and the step response is as fast as possible without overshoot. This condition is satisfied by

$$\left(\sqrt{\frac{\tau_\mu}{\tau_p}} + \sqrt{\frac{\tau_p}{\tau_\mu}}\right)^2 = 4(\mu K + 1) \tag{29}$$

Good fidelity requires that $\mu K \gg 1$. Therefore Equation 29 is satisfied when a) $\tau_\mu \gg \tau_p$ or b) $\tau_\mu \ll \tau_p$. The two solutions and the corresponding values of τ_0 (Eq. 28) are

$$\tau_\mu \gg \tau_p \qquad \tau_\mu = 4(\mu K + 1)\tau_p \qquad \tau_0 = 2\tau_p \tag{30a}$$

$$\tau_\mu \ll \tau_p \qquad \tau_\mu = \tau_p/4(\mu K + 1) \qquad \tau_0 = \tau_p/2(\mu K + 1) \tag{30b}$$

These criteria for critical damping were determined by Katz and Schwartz (12) by solving the circuit differential equations.

The 10%–90% rise time for the step response of the critically damped second-order system is equal to $3.37\tau_0$, which for each solution is

$$\tau_\mu \gg \tau_p \qquad t_r = 6.74\tau_p \tag{31a}$$

$$\tau_\mu \ll \tau_p \qquad t_r = 1.69\tau_p/(\mu K + 1) \tag{31b}$$

When microelectrodes are used to clamp cells, condition b yields a satisfactory voltage clamp that can be made with standard amplifiers. For example, in the model $\tau_p = 5$ ms and $K = 0.5$. Thus, with $\mu = 200$, t_r is 84 μs and τ_μ for critical damping is 12.4 μs.

On the other hand, condition a is usually more appropriate with axial-wire preparations. When axons are impaled by a low-resistance axial wire, the value of τ_p is very small, typically 6 μs (12). In this case, with $\mu = 100$ (and $K \approx 1$), t_r is 40 μs and the amplifier time constant for critical damping is 2.4 ms. If condition b were used with the axial-wire preparation, the amplifier time constant required for critical damping would be 15 ns. Although the predicted step response would be very fast (100 ns), this approach cannot be used because high-gain low-noise DC-coupled amplifiers cannot be made to operate over the required bandwidth. Even if they could, interactions between cables and microelectrodes and other high-frequency effects would destabilize the response. The intermediate case represented by cells with $\tau_p \approx 1$ ms can be handled by a compensation technique discussed in the chapter by Lecar and Smith.

The Bode gain plot for the critically damped case (condition b) is shown in Figure 10. The system is stable because the slope does not reach 40 dB/decade until well after the point at which the open-loop gain crosses through the 0-dB line.

Practical implications. From Equation 28 it can be seen that the damping ratio depends on μK and on the values of τ_p and τ_μ. Equation 28 simplifies to $\xi = \sqrt{R_{e2}C_m/4\mu\tau_\mu}$ at the high values of μ required for $\alpha \approx 1$. If the voltage clamp is adjusted for a critically damped step response at the beginning of an experiment and then R_{e2} decreases (R_{e2} frequently varies during passage of current), ξ will fall below unity and the step response will become underdamped. Although with microelectrodes it is more likely that R_{e2} will increase, both variations are possible. Underdamped responses are less easy to interpret than overdamped responses and may be highly undesirable (cf. the chapter by Lecar and Smith). A less fickle approach than adjusting for critical damping is to make τ_p^{-1} into a dominant pole by making τ_μ as small as possible. In the limit as $\tau_\mu \to 0$, the voltage clamp reverts to the ideal first-order system and μ can be increased until the desired rise time is achieved.

This section shows that it is best to make τ_μ as small as possible. However, when clamping real cell membranes it is often necessary to give τ_μ some significant value in order to get stability and rapid clamping. This must be done because of the nonideal frequency and phase responses

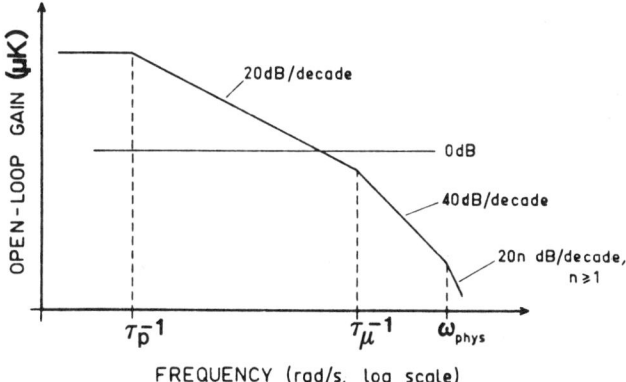

Fig. 10. Asymptotic Bode gain plot for critical damping (condition b; Eq. 30).

of real cell membranes. In this chapter we assume that an isopotential membrane can be modeled by a parallel RC circuit. At high frequencies the capacitance of this model introduces a 90° phase shift between membrane current and potential. However, even though it may be isopotential, a real membrane often has a smaller phase shift than this (24). If so, τ_μ can be used to increase the average total phase shift around the loop to approximately the 90° phase shift seen in the ideal first-order voltage clamp.

Added phase lead to compensate amplifier input time constant

One performance limitation that the experimenter cannot avoid is the limited bandwidth of the recording microelectrode. This introduces a pole at τ_{in}^{-1} in the feedback circuit that significantly deteriorates the response speed. The solutions to this limitation, in order of desirability, are to *1)* reduce the stray capacitances at the input of the unity-gain buffer (this can be achieved with low levels of bathing solution, by careful layout of input connections, and with low-input-capacitance stage as first stage of buffer); *2)* use a negative-capacitance circuit to compensate most of the lumped input capacitance; and *3)* incorporate a zero into the voltage-clamp feedback path.

No compensation circuit can fully compensate for the presence of C_{in} because of bandwidth limitations in the amplifiers used in the compensation circuit and because C_{in} is in practice made up partly of distributed capacitance. Also, all compensation circuits introduce noise. For these reasons it is far better to limit C_{in} in the first place (solution *1*) than to try to compensate for it (solutions *2* and *3*). Solution *2* is preferable to solution *3* because the voltage monitored by the buffer amplifier is made into a closer approximation of the actual intracellular voltage. Incorporating a zero at τ_z^{-1} in the feedback path is used when the first two solutions are insufficiently applied. It has been commonly used in voltage-clamp circuits (e.g., refs. 2 and 11) and its use is described next in conjunction with the description of the effects due to τ_{in}.

To analyze the effects of τ_z and τ_{in}, it is assumed that $\tau_s = \tau_\mu = 0$. The transfer function in Equation 25 simplifies to

$$\frac{V_m(s)}{V_c(s)} = \frac{\mu K(s\tau_{in} + 1)}{(s\tau_p + 1)(s\tau_{in} + 1) + \mu K(s\tau_z + 1)} \quad (32)$$

To achieve the same response as the ideal voltage clamp, C_z must be set so that $\tau_z = \tau_{in}$. Generally, however, $\tau_z \neq \tau_{in}$ and expressing Equation 32 in polynomial form yields

$$\frac{V_m(s)}{V_c(s)} = \frac{\alpha(s\tau_{in} + 1)}{s^2\tau_p\tau_{in}/(\mu K + 1) + s(\tau_p + \tau_{in} + \mu K\tau_z)/(\mu K + 1) + 1} \quad (33)$$

Comparing coefficients with the standard form of the all-pole second-order function (Eq. 27) yields the predominant time constant and the damping ratio

$$\tau_0 = \sqrt{\frac{\tau_p \tau_{in}}{\mu K + 1}} \qquad (34)$$

and

$$\xi = \sqrt{\frac{(\tau_p + \tau_{in} + \mu K \tau_z)^2}{4(\mu K + 1)\tau_p \tau_{in}}}$$

Because of the zero in the numerator in Equation 33, the step response of this second-order system may be very different from the step response of the all-pole second-order system described in the previous section. The exact response can be found from tabulated data for standard single-zero two-pole functions (5). Descriptively, the responses are as follows.

When $\tau_z = 0$ (i.e., no introduced zero), the damping ratio approaches zero as $\mu K \to \infty$; i.e., at higher gains the response shows progressively more overshoot and instability occurs. This can be seen in the Bode gain plot in Figure 11A, which crosses the 0-dB line at a slope of 40 dB/decade.

As τ_z is raised from zero, but $\tau_z^{-1} > \tau_{in}^{-1}$, the damping ratio is always larger for a given value of μK and overshoot is less likely. Stability may be achieved at high gains because the slope of the Bode gain plot may be only 20 dB/decade as the 0-dB line is crossed (Fig. 11B).

If τ_z^{-1} is set less than τ_{in}^{-1}, the slope of the Bode gain plot never exceeds 20 dB/decade (Fig. 11C) and stability is ensured. Correspondingly, $\xi > 1$ for large μK and the response settles monotonically.

It seems then that setting $\tau_z^{-1} < \tau_{in}^{-1}$ is a good procedure. However, in practice, a new limitation arises because of the poles at ω_{phys}. The dashed lines (Fig. 11C) show that if τ_z^{-1} is too small, a plateau is introduced into the Bode gain plot; this allows the poles at ω_{phys} to become effective before the 0-dB line is crossed, thereby causing instability.

The conclusion to be drawn is that too little or too much added phase lead will cause an oscillatory response. In practice, the amount of C_z added should be adjusted until the best step response is achieved. It is often better to use too little rather than too much because introducing C_z greatly increases the noise level of the clamp. This is because C_z has a low resistance to high-frequency signals, and thus it allows high-frequency noise from the microelectrode buffer amplifier into the summing junction.

It may be shown that the same Bode gain plot, and thus stability, could be achieved if τ_z were placed in the forward amplifying path (i.e.,

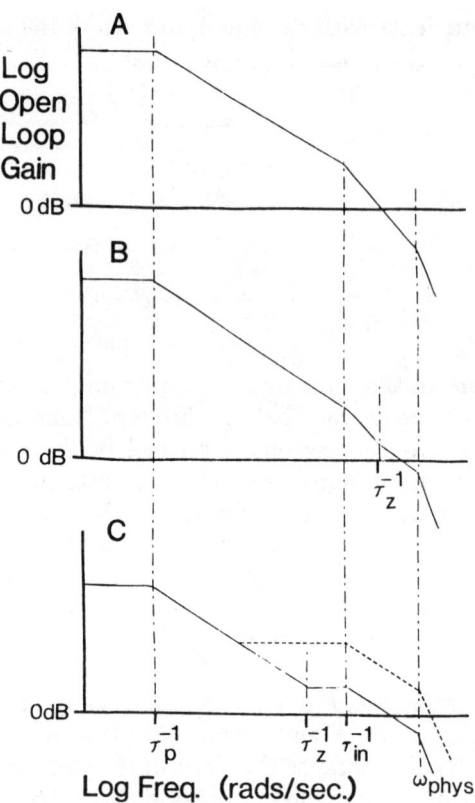

Fig. 11. Asymptotic Bode gain plots showing compensation for pole at τ_{in}^{-1} by placing zero at τ_z^{-1}. A: $\tau_z^{-1} = \infty$. System is unstable at gain shown because rate of attenuation is 40 dB/decade at unity-gain (i.e., 0 dB) crossover frequency. B: $\tau_z^{-1} > \tau_{in}^{-1}$. Rate of attenuation has been reduced to stable value but system may be highly oscillatory. C: $\tau_z^{-1} < \tau_{in}^{-1}$. Average rate of attenuation has been reduced to <20 dB/decade thus ensuring stability and overdamped response. This is true only if reduced rate of attenuation does not increase unity-gain crossover frequency to value at which poles at ω_{phys} (caused by physical realization of circuit) cause instability.

somewhere after summing junction). However, $V_m(s)/V_c(s)$ in this case has an extra term in the numerator and does not reduce to the ideal case when $\tau_z = \tau_{in}$. It is therefore not considered here. In some cases, a series of poles and zeros are added into the forward amplifying path to modify the average slope of the Bode gain plot. This technique, used to cope with more complicated membranes than the one modeled here, is discussed in the chapter by Lecar and Smith. One of the poles (often the only pole or zero) used for this purpose is the pole at τ_μ^{-1} (because of C_μ in Fig. 9A).

Irrespective of whether τ_z is placed in the forward or the feedback path, the transfer function $I_m(s)/G_m(s)$ is identical in form to the corresponding transfer function $V_m(s)/V_c(s)$.

Effects of series resistance

Origin and errors caused. The series resistance shown in Figure 9A may arise from a number of sources. One of these is the bulk resistance (R_{ss}) of the cytoplasm, membrane infolds, and extracellular solution. Normally R_{ss} would be insignificant compared with R_m; however, in some circumstances it may not remain so. For example, after depolarization to +20 mV, R_m in some snail cells falls from rest values of 3–5 MΩ (14) to <50 kΩ. Under these circumstances R_{ss}, which is around 8–12 kΩ, can have a significant effect on clamp dynamics and accuracy.

Another component of R_s is the coupling resistance (R_x) that exists in the vicinity of the microelectrode tips. Although commonly used, the term *coupling* in this connection is not a good one because R_x is not the resistance linking the two microelectrodes: it is the resistance to ground seen by one microelectrode when current flows from the tip of the other (see Fig. 13). When current flows from the tip of ME_2, there is a considerable voltage drop in the first few micrometers of fluid surrounding the tip because of the high current density occurring before the current spreads out. Some of the voltage drop occurs across the resistance (R in Fig. 13) linking the two microelectrodes and some across R_x. The proportion attributable to R_x depends on the separation of the microelectrode tips.

The voltage drop across R is indistinguishable from the voltage drop across ME_2 and has no effect on the voltage clamp other than in the determination of K. However, the voltage drop across R_x, like the voltage drop across R_{ss}, is recorded by ME_1 and interpreted by the voltage-clamp circuit as being part of the membrane potential.

When two physically separate microelectrodes are used for voltage clamping, the tips are usually far enough apart (several micrometers) for R_x to be too small to have a significant effect on the clamp response. However, if the two microelectrodes are the halves of double-barreled glass tubing (theta glass), the value of R_x may reach several hundred kilohms. Thus this type of glass is very unsuitable for two-electrode voltage clamping.

If R_s is large compared with R_m, the voltage-clamp circuit mainly clamps the voltage across R_s. The step response of $V_{mm}(s)/V_c(s)$ is very fast even at quite low gains because there is negligible capacitance to be charged. The step response of the real membrane potential [$V_m(s)/V_c(s)$], which cannot be measured, is much slower and attenuated. Also, membrane current responses to changes in G_a do not correspond to $V_{mm}(s)/V_c(s)$ because the requirement that G_a be in parallel with the whole of the membrane across which V_{mm} is measured is not met. In fact $I_m(s)/G_a(s)$ has the same poor step response as the unmeasurable step response of V_m.

Even though the measurement of V_m is in error, the measurement of I_m is not. Because R_s and the membrane are in series with each other and with the current-measurement circuit, the measurement of I_m is always the true membrane current.

In many experiments (e.g., when synaptic currents are recorded) $R_m \gg R_s$ (where $R_s = R_{ss} + R_x$) and therefore R_s has an insignificant effect on the voltage-clamp response. In other experiments (e.g., when active currents are recorded during large depolarizations) R_s cannot be ignored and the following discussions apply.

Stability. A CLTF of the same form as Equation 32 arises if τ_z is set equal to τ_{in} but τ_μ and τ_s have nonzero values, where $\tau_s = (R_m \| R_s)C_m$. In this case the transfer function in Equation 25 reduces to

$$\frac{V_{mm}(s)}{V_c(s)} = \frac{\mu K(s\tau_s + 1)}{(s\tau_\mu + 1)(s\tau_p + 1) + \mu K(s\tau_s + 1)} \quad (35)$$

The value of τ_s is not under the direct control of the experimenter, but the value of τ_μ can be changed. Therefore τ_μ can be set equal to τ_s so that Equation 35 reduces to the ideal transfer function.

If τ_μ does not equal τ_s, the effect on stability is similar to the effect on stability of τ_{in} and τ_z as discussed in the previous section; i.e., a wide range of values of τ_μ/τ_s exists for which stability of the system is assured (unless the poles at ω_{phys} come into effect). A similar situation would arise if τ_μ were zero but τ_{in} were present; i.e., stability would be improved because the open-loop zero at τ_s would, for a range of values, tend to compensate the open-loop pole τ_{in}.

In summary, if the poles at ω_{phys} can be ignored, the open-loop zero from R_s may improve stability of a voltage clamp that would otherwise be unstable at high gains because of the open-loop pole at τ_μ or τ_{in}.

Compensation. The possible stabilizing of the clamp by R_s is offset by the fact that ME_1 measures the voltage (V_s) across R_s and across the membrane. Thus, even though the clamp may be more stable than it would have been without R_s, the clamp is less useful because at high membrane current levels the membrane potential differs from the recorded potential (V_1) by V_s ($= I_m R_s$).

To compensate for the measurement error introduced by R_s, it is necessary to subtract V_s from V_1. Because V_s cannot be measured directly for physical reasons, it is derived electronically from $I_m R_s$. The membrane current is continuously available from the current-measurement circuit, and R_s can be estimated before voltage clamping from the size of the initial step change in potential recorded by one intracellular microelectrode when a current step is injected via the other microelectrode. [A more accurate technique for measuring R_s has been described for axon voltage clamps (15). In this technique R_s is calculated from the membrane impedance measured over a wide frequency range. This determination of R_s is more difficult to apply than the current-step technique.]

Figure 12A shows the block diagram with the compensation function included. The forward-path transfer function has been represented by $X(s)$. The phase-lead network ($s\tau_z + 1$), sometimes present in the feedback path, has been left out. Instead, τ_{in} represents the net microelectrode time constant effective after all neutralization and compensation measures have been taken. The DC gain (μ_s) of the R_s-compensation feedback pathway can be set in the range 0–1. To achieve complete substraction of V_s at low frequencies, μ_s should be unity. The term $(s\tau_f + 1)^{-1}$ is explained in the next section, as is the dotted pathway element.

High-frequency considerations. If implemented as shown, the clamp circuit is sure to oscillate. The measured membrane current is defined by the circuit (see Fig. 9A) to be

$$I_m(s) = \frac{V_{mm}(s)}{Z_m(s) + R_s}$$

Multiplication of I_m by $\mu_s R_s$ yields a voltage (V_{ss}) that is equal to $\mu_s V_s$. That is

$$V_{ss}(s) = \frac{\mu_s R_s}{R_m + R_s} \frac{(s\tau_m + 1)}{(s\tau_s + 1)} V_{mm}(s) \quad (36)$$

$$= H_{ss}(s) V_{mm}(s)$$

where $H_{ss}(s)$ is the frequency-dependent gain of the compensation feedback pathway. At frequencies $>\tau_s^{-1}$, $H_{ss}(s)$ becomes equal to μ_s.

Equation 36 shows that $V_{ss}(s)$ is a function of the clamped potential $V_{mm}(s)$. Instead of being shown in Figure 12A as having been derived from $I_m R_s$, $V_{ss}(s)$ could equivalently be shown as having been derived from $V_{mm}(s)$ by $H_{ss}(s)$ as shown in the dotted pathway. The frequency dependence of $H_{ss}(s)$ is shown in Figure 12B.

Because there are two inverting summers in the compensation feedback pathway, the feedback of $V_{mm}(s)$ via this pathway is positive. As long as the magnitude of $H_{ss}(s)$ is less than the gain $[H_f(s)]$ of the usual feedback pathway, the net feedback of $V_{mm}(s)$ remains negative and the circuit does not oscillate. At low frequencies ($<\tau_m^{-1}$), this condition is satisfied because $H_f(0) = 1$, whereas $H_{ss}(0) \ll 1$ (because usually $R_s \ll R_m$). At high frequencies ($>\tau_s^{-1}$), $H_{ss}(s)$ asymptotes to μ_s (Eq. 36; see Fig. 12B). Thus at these frequencies the voltage-clamp circuit is no longer a unity-feedback circuit. For values of μ_s less than unity the circuit may remain

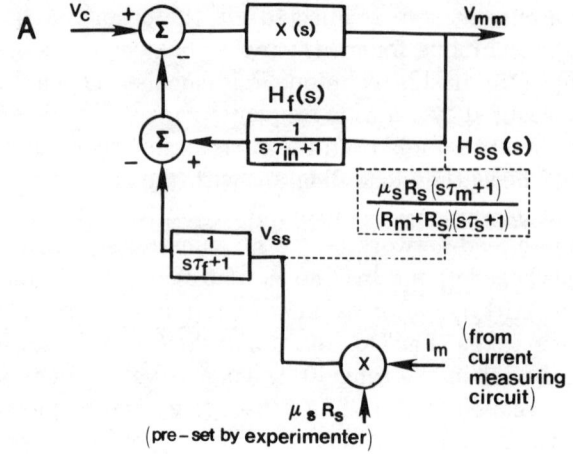

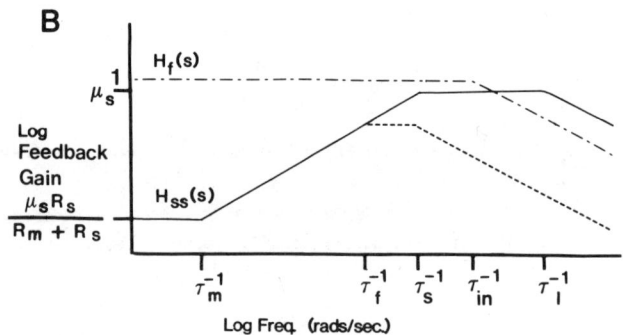

Fig. 12. Compensation for error due to resistance (R_s) in series with membrane. A: basic clamp circuit consisting of forward-path transfer function [$X(s)$] and feedback-path transfer function [$(s\tau_{in} + 1)^{-1}$] has been supplemented by addition of circuit to subtract V_{ss}, calculated and scaled value of V_s (voltage across R_s); V_{ss} is calculated by multiplying I_m (recorded by current-measuring circuit) by $\mu_s R_s$, where μ_s is scaling factor (gain) in range 0–1. Transfer function enclosed in *dotted box* is mathematical equivalent of multiplying circuit and shows that V_{ss} depends on clamped potential [$V_{mm}(s)$]. Transfer function [$(s\tau_f + 1)^{-1}$] is necessary for stability when μ_s approaches unity. B: frequency response of feedback pathways in A; $H_f(s)$, usual feedback pathway [$(s\tau_{in} + 1)^{-1}$]; $H_{ss}(s)$, compensation feedback pathway shown in *dotted box* in A; τ_I^{-1}, bandwidth of current-measurement circuit. At frequencies $>\tau_{in}^{-1}$, $H_f(s)$ becomes $<H_{ss}(s)$ and net feedback becomes positive. To prevent this, pole is introduced at τ_f^{-1} to limit high-frequency gain of $H_{ss}(s)$ (*dotted line*).

stable; however, for values of μ_s approaching unity the step response has a lot of overshoot and there is excessive high-frequency noise. In practice, even if $\mu_s < 1$, the circuit still oscillates because the bandwidth (τ_I^{-1}) of the current-measurement circuit is usually $>\tau_{in}^{-1}$. Therefore, at high frequencies $H_f(s)$ drops below $H_{ss}(s)$ and the net feedback becomes positive (Fig. 12B). Typically, μ_s values in the range 0.6–0.9 are the maximum that can be achieved (11, 14).

To fully compensate for $V_s(s)$ at low and medium frequencies, it is necessary to use $\mu_s = 1$ and to attenuate $V_{ss}(s)$ at high frequencies by a pole at a frequency $<\tau_s^{-1}$ or τ_{in}^{-1}, whichever is lower. The required pole is $(s\tau_f + 1)^{-1}$ shown in Figure 12A. The attenuation provided by the pole ensures that $H_{ss}(s)$ is always significantly smaller than $H_f(s)$ so that the net feedback remains negative at all frequencies (Fig. 12B, dotted line).

A similar compensation scheme was described by Sigworth (19) for voltage clamping myelinated nerve. An alternative technique has recently been described by Moore et al. (15) for voltage clamping unmyelinated axons. In this method an active bridge circuit is used to separate the capacitive membrane currents from the ionic membrane currents, and the wide-bandwidth compensation for R_s is made on the ionic currents only. Thus the method yields full compensation for that component of V_s caused by ionic currents but no compensation for the component of V_s caused by capacitive currents. The compensation is stable because the total feedback due to all currents (i.e., ionic and capacitive) is always less than unity.

Interelectrode capacitance

An extremely important component of most voltage clamps is the interelectrode or coupling capacitance C_x (sometimes referred to as cross capacitance; Fig. 13), which links the two microelectrodes through the air space and bath fluid between them or, for theta glass and fabricated microelectrode assemblies, through the common glass wall (9).

The value of C_x linking two microelectrodes ranges from <0.01 pF for separate microelectrodes shielded from each other to many tens of picofarads for theta glass. Even at the lowest range of values, C_x adversely affects the dynamic response of the voltage clamp or perhaps causes it to oscillate.

The interelectrode capacitance introduces a second feedback pathway that bypasses the cell, going directly from the output of the clamp amplifier to the input of the voltage-recording amplifier. The transfer function of this coupling pathway, given that $C_x \ll C_m$ and $R_s \ll R_{el}$, is

$$\frac{V_1(s)}{V_o(s)} = \frac{s\tau_x}{s\tau_{IN} + 1} \tag{37}$$

where $\tau_x = R_{el}C_x$ and $\tau_{IN} = R_{el}(C_{in} + C_x)$. The transfer-function block diagram incorporating the coupling pathway is illustrated in Figure 14

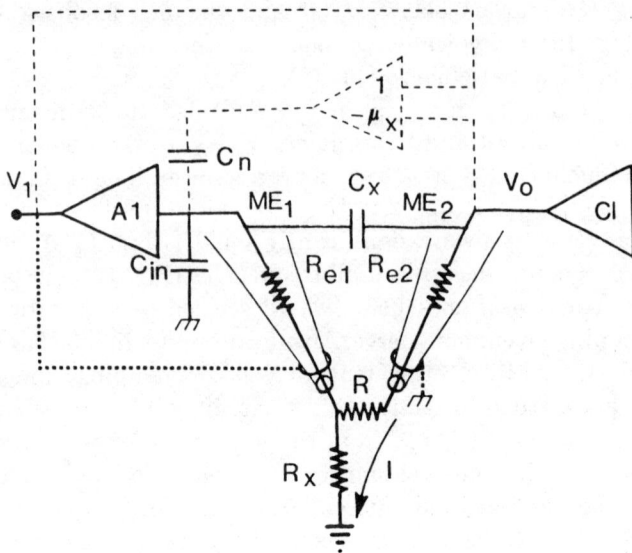

Fig. 13. Sources of direct coupling between two microelectrodes. R_x, coupling resistance; C_x, coupling (or "cross") capacitance; A1, headstage amplifier. *Dotted lines* show shielding and guarding arrangements used to reduce C_x; *dashed lines* show neutralization circuit used to reduce C_x. Unity gain feedback from ouput of A1 to input of compensating amplifier bootstraps C_n (neutralization capacitance) so that signals recorded by ME_1 are not slowed down by presence of C_n.

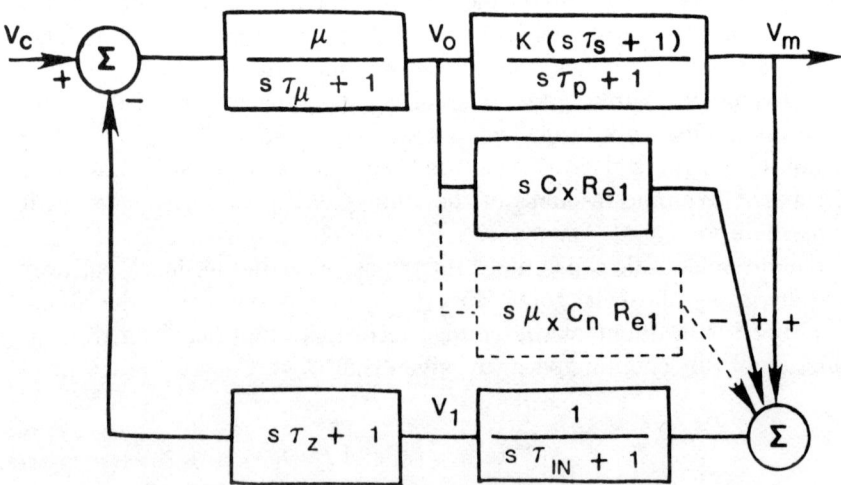

Fig. 14. Transfer-function block diagram incorporating second pathway caused by coupling capacitance (C_x). *Dashed pathway* is used to electronically compensate for C_x.

(for present, ignore dashed pathway). Note that the presence of C_x has decreased the frequency of the pole in the feedback pathway from τ_{in}^{-1} to τ_{IN}^{-1}.

The open-loop gain of this second circuit can be written down from the transfer-function block diagram. That is

$$\text{OLTF}(s) = \frac{\mu s \tau_x (s \tau_z + 1)}{(s \tau_\mu + 1)(s \tau_{IN} + 1)} \tag{38}$$

The Bode gain plot is shown in Figure 15. There is no dominant pole. Instead the gain curve is either increasing or flat for all frequencies until ω_{phys} is reached, at which point the multiple poles from the physical realization of the circuit become effective. If at this frequency the open-loop gain has risen above unity, the circuit will oscillate because the excess phase shift at ω_{phys} will exceed $-180°$. To prevent this possibility, it is necessary to make C_x exceedingly small. As the frequency increases, the largest magnitude of open-loop gain achieved is $\mu \tau_x \tau_z / \tau_\mu \tau_{IN}$. Thus the condition for stability is that $\tau_x < \tau_\mu \tau_{IN} / \mu \tau_z$. For the values used already (see *Finite bandwidth of electronic amplifying circuit*, p. 67; $\mu = 200$ and $\tau_\mu = 12.4$ μs) and taking $\tau_z = 6.2$ μs, the condition for stability is $\tau_x < \tau_{IN}/100$, which from the definition of τ_x and τ_{IN} means that $C_x < C_{in}/99$. Interestingly the destabilizing effect of the interelectrode capacitance is worsened when C_{in} is kept small.

The best way to keep C_x very small is by placing a grounded shield around ME_2 or between the two microelectrodes. The minimum requirement for the shield is that it should extend from the point at which the microelectrode is connected to the ME_2 headstage down to within 1 mm

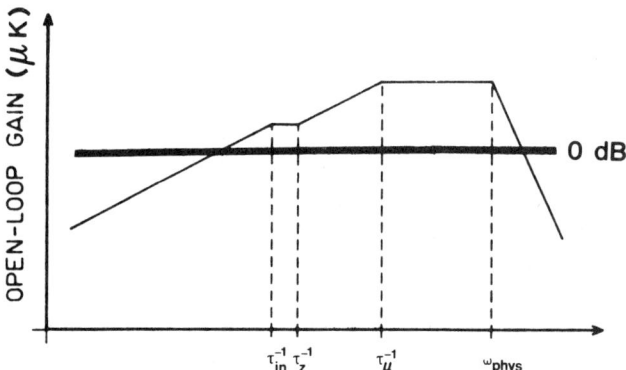

Fig. 15. Bode gain plot of second pathway caused by coupling capacitance (C_x), which introduces zero in response and causes initial 20-dB/decade increase in open-loop gain. If C_x is too large, open-loop gain will exceed 0 dB (as shown) before other poles in system become effective and clamp will oscillate.

of the bathing solution. More extensive shielding may be achieved by insulating the shield and allowing it to dip into the solution. Capacitive coupling between the immersed portions of the microelectrodes is strongly diminished by the presence of the grounded solution, but it is not entirely eliminated because the salt solution is not a zero-resistance fluid. In some cases it may be necessary to paint or otherwise deposit an insulated metallic shield onto the outer wall of ME_2 extending as close to the tip as possible. In rare circumstances the same procedure may also be required for ME_1, but in this case the shield is guarded or driven by a voltage equal to the input signal (see Fig. 13, dotted line). The case of the ME_2 headstage should also be grounded (for shielding details, see the chapter by Sachs). In addition, the angle between ME_1 and ME_2 should be 90° or more if possible. These measures may reduce C_{in} to <0.01 pF. Nevertheless, for large values of μ, the transient response of the voltage-clamp circuit may be impaired by the presence of C_x. It can be improved by electronically neutralizing the residual value of C_x (9). This can be achieved by applying an appropriately scaled inverted version of V_o to a capacitor (C_n) connected to the input of A1 as shown by the dashed pathways in Figures 13 and 14. The output of amplifier A1 is

$$V_1 = \frac{V_m}{s\tau_{IN} + 1} + \frac{sR_{e1}(C_x - \mu_x C_n)V_o}{s\tau_{IN} + 1} \tag{39}$$

where μ_x is the scaling factor of the neutralization amplifier. Therefore the effective interelectrode capacitance is

$$C_{eff} = C_x - \mu_x C_n \tag{40}$$

Thus the influence of C_x can be nullified by setting

$$\mu_x = C_x/C_n \tag{41}$$

If μ_x is made larger than this value, the net feedback in the second pathway becomes positive and the circuit becomes unstable. As μ_x approaches this critical value, the high-frequency noise will increase because C_n has a low resistance to high-frequency signals.

It is possible to set μ_x before switching to voltage-clamp mode after the microelectrodes are in position in the cell by putting a square voltage waveform onto ME_2 and adjusting μ_x to minimize the pickup on ME_1 (i.e., on the V_1 record). The correct setting of μ_x is independent of R_{e2} and should not need to be readjusted until the microelectrodes are moved.

In practice, C_x cannot be completely neutralized by this technique because it is a distributed capacitance not well approximated by a single capacitor. Therefore the neutralization technique should be used in addition to the techniques described for minimizing C_x, not instead of them. If the residual value of C_x is large, such as occurs if the two barrels of theta glass are used for voltage clamping, the neutralization will probably never be sufficient to achieve good clamp performance.

Changes in bath potential

Several factors can cause the potential of the bathing solution (V_B) to change. Extraneous factors include line-frequency pickup (hum) and changes in the chemical composition of the bathing solution or in its temperature. An inherent component of V_B is the IR voltage drop caused by current flow through the bulk resistance of the bathing solution (R_B) and the grounding electrode. This R_B is the extracellular component of R_{ss} (see *Origin and errors caused*, p. 73). The value of R_B would normally be insignificant compared with R_m but may become significant if the ground electrode has high resistance or if a virtual ground (see the chapter by Finkel) is used and it is slow. The value of V_B from the extraneous factors alone is designated V_B'.

The intracellular voltage-recording microelectrode records V_B and V_m. The voltage-clamp circuit then tries to clamp $V_m + V_B$ to the command potential. Because clamping V_m alone is desired, V_B must be eliminated from the signal that the voltage-clamp circuit tries to clamp. This can be done in two ways.

Differential-input voltage-clamp circuit. The usual technique to remove V_B is to use a third microelectrode (ME_3) to record V_B close to the cell and to subtract V_B from the voltage recorded by ME_1 with a differential amplifier. This is illustrated in Figure 16, which is based on the ideal voltage clamp shown in Figure 4. The low-frequency and midfrequency (i.e., $<\tau_0^{-1}$) circuit equations are

$$V_o = \mu(V_c - V_m)$$
$$V_m = K(V_o - V_B)$$

where

$$K = R_m/(R_m + R_{e2} + R_B)$$

Combining these equations to eliminate V_o yields

$$V_m = \alpha V_c - \frac{\alpha}{\mu} V_B \tag{42}$$

From this equation it can be seen that for the differential-input circuit the clamped membrane potential depends to some extent on the bath potential. However, this dependence becomes exceedingly small at large values of μ.

Voltage-clamp circuit referenced to bath potential. All of the dependence of the clamped membrane potential on the bath potential can be elimi-

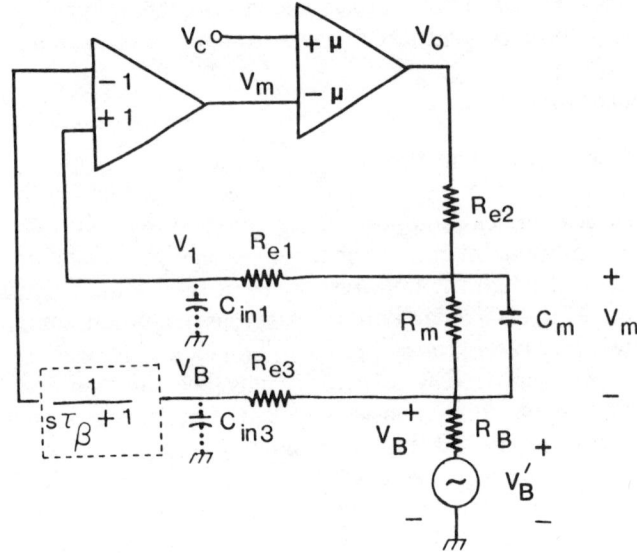

Fig. 16. Differential-input voltage-clamp circuit. R_{e3}, resistance of ME_3 [extracellular microelectrode used to record bath potential (V_B)]. At low frequencies, clamped membrane potential is almost independent of V_B (see Eq. 42). At high frequencies, subtraction of V_B causes instability because recording bandwidth of ME_3 [= $(R_{e3}C_{in3})^{-1}$] usually exceeds recording bandwidth of ME_1 [= $(R_{e1}C_{in1})^{-1}$]. To prevent instability, bandwidth for recording V_B is limited by low-pass filter shown in *dotted box*.

nated by slightly modifying the previous circuit (Fig. 17). The low-frequency and midfrequency circuit equations are

$$V_o = \mu(V_c - V_m) + V_B$$

and

$$V_m = K(V_o - V_B)$$

Combining these equations to eliminate V_o yields

$$V_m = \alpha V_c \tag{43}$$

which shows that for the V_B-referenced circuit the clamped membrane potential is completely independent of the bath potential.

High-frequency considerations. At high frequencies the changes in V_B principally reflect the voltage drops across R_B caused by I_m. The elimination of V_B involves the positive feedback of the exact amount of V_B that will cancel the amount of V_B already present in the negative-feedback circuit by virtue of its being a component of the voltage recorded by ME_1. At high frequencies the cancellation of V_B is no longer exact because of the difference in the recording bandwidths of ME_1 and ME_3.

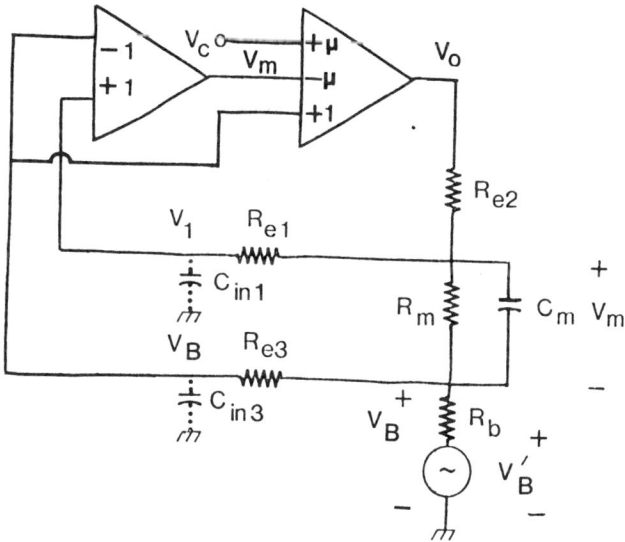

Fig. 17. Bath potential–referenced voltage-clamp circuit is almost the same as differential-input voltage-clamp circuit. Only difference is unity addition of V_B into summing amplifier. Thus at low frequencies clamped membrane potential is independent of V_B (Eq. 43). For clarity, low-pass filter is not shown.

Typically, ME_3 is a broken-tipped or agar-filled electrode with very low resistance; thus its recording bandwidth is normally considerably higher than that of ME_1. Therefore at high frequencies ($>\tau_{in}^{-1}$) the amount of V_B in the negative-feedback circuit as measured by ME_1 decreases, leaving a net positive feedback of V_B (measured by ME_3) that destabilizes the circuit.

To prevent this destabilization, the bandwidth of the bath-potential recording circuit must be limited by a low-pass filter with a cutoff frequency $<\tau_{in}^{-1}$. The required filter is shown in the dotted box in Figure 16. The filter time constant (τ_B) is typically 100 μs or more.

General Considerations

Sensitivity to component variation

In this section the dependence of the stability and rise time of the ideal voltage clamp on the values of R_m, C_m, and R_{e2} is considered. To compare the effects of changes in each component, the voltage-clamp gain is adjusted to maintain $\alpha = 0.999$.

Resistance of current-passing microelectrode. The resistance of the current-passing microelectrode affects both τ_p and K. For $R_{e2} = 10$ MΩ and $\alpha = 0.999$ in the cell model, $\tau_p = 5$ ms, $K = 0.5$, and $\mu = 2,000$. If R_{e2}

is decreased from 10 MΩ to a value much less than R_m (say 1 MΩ), K increases from 0.5 to 0.91 whereas τ_p decreases from 5 ms to 0.91 ms. To keep $\alpha = 0.999$, μ is decreased from 2,000 to 1,100. The Bode gain plots for the initial and subsequent situations are shown by the continuous and dotted lines, respectively, in Figure 18. In this particular example the clamp is unstable for $R_{e2} = 1$ MΩ because the dominant pole at τ_p^{-1} has been pushed closer to ω_{phys}, to the extent that the slope of the Bode gain plot exceeds 40 dB/decade as the 0-dB line is crossed. If by careful electronic design ω_{phys} had been at a higher frequency or if the cell under investigation had had a larger value of C_m so that τ_p was larger, stability may have been preserved. In that case the significant advantage of reducing R_{e2} would have been a decrease in the step-response rise time. From Equation 8 it can be seen that for a 10-fold decrease in R_{e2} and a simultaneous reduction in μ from 2,000 to 1,100, t_r decreases to 0.18 of its earlier value.

On the other hand, increasing R_{e2} until it is much larger than R_m has less effect on stability and rise time. The maximum increase in τ_p that will result as R_{e2} increases toward infinity is a doubling from 5 ms to 10 ms. Thus any improvement in stability (caused by shifting pole at τ_p^{-1} away from ω_{phys}) is marginal. If μK is adjusted to remain constant (to preserve $\alpha = 0.999$; Eq. 4), Equation 5 shows that τ_0 (and thus t_r) is

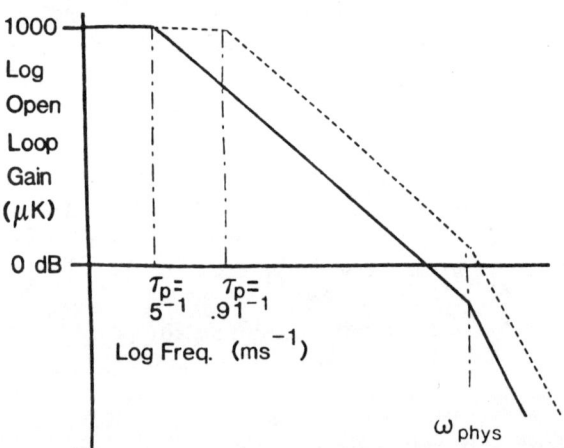

Fig. 18. Bode gain plot showing effect on stability of decreasing value of R_{e2}. Initially (*solid line*) $R_{e2} = 10$ MΩ, $K = 0.5$, $\tau_p = 5$ ms, $\mu = 2,000$, and $\alpha = 0.999$. Subsequently (*dashed line*) R_{e2} is decreased to 1 MΩ, resulting in $K = 0.91$ and $\tau_p = 0.91$ ms; μ is decreased from 2,000 to 1,100 to keep $\alpha = 0.999$. In this example, clamp is unstable with $R_{e2} = 1$ MΩ because poles at ω_{phys} cause rate of attenuation of open-loop gain to exceed 40 dB/decade at unity-gain crossover frequency. If ω_{phys} had higher value, clamp would remain stable and rise time would significantly decrease to 0.18 of initial value.

proportional to τ_μ. Thus t_r can at most double from the value at which $R_{e2} = R_m$ as R_{e2} is made very large. Against the stability improvement caused by increasing τ_p is the possibility that the clamp-amplifier time constant (τ_μ) might become large enough to worsen the stability. The reason why τ_μ might increase from its original value stems from the inverse relationship between gain and bandwidth of most practical amplifiers. The gain must be increased to compensate for the decrease in K (as R_{e2} is made larger) and the increase in μ might be accompanied by an increase in τ_μ (i.e., decrease in bandwidth).

Thus there is no definite policy for selecting a value of R_{e2}. If $R_{e2} \gg R_m$, stability may be compromised because of the risk that for the large values of μ required for the desired fidelity the clamp amplifier bandwidth might be reduced. Also, ME$_2$ may not be able to pass the current required for charging C_m. If $R_{e2} \ll R_m$, stability may be worsened because the dominant pole at τ_p^{-1} is shifted toward ω_{phys}. On the other hand, the rise time is faster, which means that $R_{e2} \ll R_m$ is the best selection if maintaining stability is not a problem.

It should be noted that in an axial-wire preparation R_{e2} may diminish to zero without affecting stability because the dominant pole is at τ_μ^{-1}, which is independently set in the electronics.

Membrane capacitance. Altering the value of C_m has no effect on K or therefore on α. However, C_m affects τ_p and thus the stability and rise time. For large values of C_m, τ_p^{-1} is small and thus stability is improved because the dominant pole is moved further away from ω_{phys}. However, at constant α, t_r is increased because it depends linearly on C_m (Eq. 8). In practice, cells with large C_m values are easy to voltage clamp at high fidelity ($\alpha \to 1$) and small t_r because the stability afforded by the low frequency of the dominant pole allows very high values of μ to be used.

Membrane resistance. The value of R_m affects τ_p and K. If μ is adjusted to keep $\alpha = 0.999$ for each R_m, the effect of changes in the R_m value on stability and rise time is similar to the effect of changes in the R_{e2} value. Note that this circumstance is different from the one described in *Membrane voltage responses to changes in command potential* (p. 56), where it was shown that for fixed μ (as long as $\mu K \gg 1$) R_m is not a factor affecting t_r. In addition, for fixed μ the value of R_m does not affect stability.

Slowing the rise time of command potential

In some voltage-clamp experiments, the investigators have eliminated overshoot in V_m after a step change in V_c by using a low-pass filter to slow the rise time of V_c (10). This low-pass filter does not affect stability because it is outside the negative-feedback loop. The modified transfer-function diagram of the ideal voltage clamp is shown in Figure 19. The

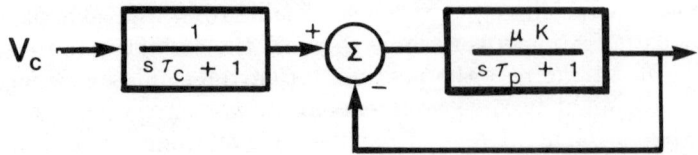

Fig. 19. Transfer-function diagram of ideal voltage clamp modified by addition of low-pass filter to slow rise time of V_c. When V_c is constant, low-pass input filter has no effect on circuit. Thus dynamic response of $V_m(s)/V_c(s)$ will not be identical to dynamic response of $I_m(s)/G_a(s)$.

effect of the filter is to modify the original transfer function given in Equation 5 by multiplying it by the filter transfer function to yield

$$\frac{V_m(s)}{V_c(s)} = \frac{1}{(s\tau_c + 1)} \frac{\alpha}{(s\tau_0 + 1)} \qquad (44)$$

where τ_c^{-1} is the cutoff frequency of the low-pass filter.

This circuit modification is not advisable because it destroys the equivalence between $V_m(s)/V_c(s)$ and the other transfer functions being investigated, $I_m(s)/G_a(s)$ and $I_m(s)/E_a(s)$. This is because, when $V_c(s)$ is constant, the input to the summing junction is constant and has no effect on the transient response of the feedback circuit arising from changes in the cell; i.e., $I_m(s)/G_a(s)$ and $I_m(s)/E_a(s)$ are not affected by the input filter and remain unmodified from the expression given in Equation 16.

If the responses require low-pass filtering, the low-pass filters should be placed in the output pathways so that all transfer functions are modified equally.

In some experiments there are practical considerations that make it advisable to slow the rise of V_c, notwithstanding the loss of equivalence between the various transfer functions. For example, when a three-electrode voltage clamp is established in muscle, the tip of the current-passing microelectrode is several hundred micrometers from the tip of the voltage-recording microelectrode. If the potential at the tip of the voltage-recording microelectrode closely follows V_c and the rise of V_c is not slowed, the large current transient required at the tip of the current-passing microelectrode might damage the cell (1).

Background noise under voltage clamp

In addition to membrane conductance fluctuations caused by activatable mechanisms, a number of other processes contribute to the fluctuations measured during voltage clamping.

Microelectrode noise. All real resistors produce thermal noise because of the random motion of current carriers. This thermal noise has a constant voltage spectral density equal to $\sqrt{4kTR}$ V/$\sqrt{Hz}$, where k is the Boltzmann constant, T is absolute temperature, R is the value of the resistor, and Hz is the system bandwidth (4, 16). In addition to the irreducible thermal noise associated with their resistance, microelectrodes also exhibit 1/f noise (see ref. 13). This noise has a voltage spectral density that diminishes with frequency. Its magnitude increases with the square of the voltage across the microelectrode and with the ionic gradient from inside to outside the microelectrode. At some low frequency, which depends on experimental parameters, the magnitude of the thermal noise begins to exceed the magnitude of the 1/f noise. Typically, above 10–100 Hz the thermal noise becomes the larger. When comparing the total noise in the microelectrode and membrane resistances in this paper, it is assumed that the total noise is similar for similar resistances.

Voltage-recording microelectrode. The combined thermal and 1/f noise of the voltage-recording circuit is an extra signal source (e_{n1}) in series with the feedback path. The transfer-function diagram is shown in Figure 20. This noise acts as an extra component of the command signal so that the clamped membrane potential is (by extending Eq. 5)

$$V_m(s) = \frac{\alpha}{s\tau_0 + 1} V_c(s) - \frac{\alpha}{s\tau_0 + 1} e_{n1}(s) \qquad (45)$$

It can be seen that the clamp circuit forces the true membrane potential to closely follow e_{n1}. The potential recorded by the microelectrode amplifier is related to V_m by

$$V_1 = V_m + e_{n1} \qquad (46)$$

Thus

$$V_1(s) = \frac{\alpha}{s\tau_0 + 1} V_c(s) + \left(1 - \frac{\alpha}{s\tau_0 + 1}\right) e_{n1}(s) \qquad (47)$$

and (because usually $\alpha \to 1$) the recorded membrane potential at low-frequency and midfrequency does not show the real effect of the microelectrode noise. To the investigator the microelectrode noise present before the voltage clamping appears to be clamped out, whereas in fact

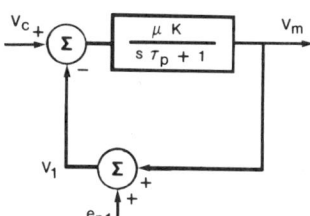

Fig. 20. Transfer-function block diagram of ideal voltage clamp showing position of noise voltage source (e_{n1}) caused by thermal noise and 1/f noise generated in ME_1.

it has been almost wholly imposed across the membrane. The only way to reduce this error is to reduce e_{n1} by using as low a value of R_{e1} as possible.

Even though the voltage noise imposed on the membrane potential is masked, the current noise caused by e_{n1} is not; $I_m(s) = V_m(s)/Z_m(s)$, which from Equation 45 yields

$$I_m(s) = \frac{\alpha}{Z_m(s)(s\tau_0 + 1)} V_c(s) - \frac{\alpha}{Z_m(s)(s\tau_0 + 1)} e_{n1}(s) \quad (48)$$

The second term in this equation is the current noise (i_{n1}) caused by e_{n1}. For the parallel cell model used so far

$$i_{n1}(s) = \frac{\alpha(s\tau_m + 1)}{R_m(s\tau_0 + 1)} e_{n1}(s) \quad (49)$$

The amplitude of i_{n1} increases at 20 dB/decade for frequencies $>\tau_m^{-1}$ and eventually flattens off at frequencies $>\tau_0^{-1}$. The low-frequency amplitude depends only on α/R_m, whereas the frequency at which the noise begins to increase depends on C_m and R_m. For intermediate frequencies at which $s\tau_m \gg 1$ and $s\tau_0 \ll 1$, the amplitude of the current noise is

$$i_{n1} = \alpha s C_m e_{n1} \quad (50)$$

Thus it can be seen that the major part of the membrane current noise caused by e_{n1} is directly proportional to the size of the membrane capacitance and to the frequency.

Current-passing microelectrode. Let ME_2 be represented by a noiseless resistor (R_{e2}) in series with a noise voltage source (e_{n2}). In the equivalent-circuit representation, R_{eq} can be represented by an ideal resistor [$R_{eq} = R_{e2}/(\mu + 1)$] in series with a noise voltage source [$e_{n,eq} = e_{n2}/(\mu + 1)$]. The membrane current caused by $e_{n,eq}$ is

$$i_{n2} = \frac{e_{n,eq}}{R_{eq} + Z_m} \quad (51)$$

$$= \frac{e_{n2}(s\tau_m + 1)}{(\mu + 1)(R_m + R_{eq})(s\tau_0 + 1)}$$

Comparison of this equation with Equation 49 indicates that, if R_{e1} and R_{e2} have similar values (so that e_{n1} and e_{n2} are similar), the membrane current noise caused by R_{e2} is $\approx 1/(\mu + 1)$ times the membrane current noise caused by R_{e1}. Therefore in normal circumstances the noise caused by R_{e2} is negligible compared with the noise caused by R_{e1}.

Membrane resistance noise. To calculate the membrane current noise caused by the noise generated in R_m, it is more convenient to represent the noise voltage source ($e_{n,m}$) as a noise current source equal to $e_{n,m}/R_m$

in parallel with a noiseless resistor of value R_m. The noise current divides between the membrane and R_{eq}, but because the amount flowing in the membrane flows wholly within the membrane, the only amount measured is the amount flowing into R_{eq} (Fig. 21). This amount is

$$i_{n,m} = \frac{e_{n,m} Z_m}{R_m (Z_m + R_{eq})} \qquad (52)$$

$$= \frac{e_{n,m}}{(R_m + R_{eq})(s\tau_0 + 1)}$$

Comparison of this equation with Equation 49 indicates that, if R_{e1} and R_m have similar values and for the usual conditions $\alpha \to 1$ and $R_{eq} \ll R_m$, the contribution to membrane noise from the two resistors is similar at very low frequencies; however, although the contribution from R_{e1} increases at frequencies $>\tau_m^{-1}$, the contribution from R_m remains constant before beginning to fall at the much higher frequency τ_0^{-1}. Thus summation of the noise over the whole of the useful clamping bandwidth (DC to τ_0^{-1}) shows that the total noise from R_m is much less than from R_{e1}. Other sources of current fluctuations in the membrane, [e.g., shot noise (21) and active membrane transport processes (22)] are insignificant compared with i_{n1} in the useful clamping bandwith.

Instrumentation noise. All practical operational amplifiers generate noise. In addition the feedback resistors used to convert the operational amplifier into a practical amplifying configuration also introduce noise. In most circumstances the noise contribution of the voltage-clamp electronics is less than the thermal noise contribution of R_{e1}. A calculation of the magnitude of the instrumentation noise is discussed by Fishman et al. [(8); see also ref. 4]. As a general rule, the input stage of the voltage-clamp amplifier should be high gain (10 or more) to minimize the noise contribution of later stages.

In summary, the main source of noise in the clamp current is caused by the voltage-clamp circuit clamping the noise generated in ME_1. The dependence of the current noise on this source is greatest at high frequencies (Eqs. 49 and 50). At these frequencies the main source of noise in ME_1 is the thermal noise. The magnitude of the noise is worst in cells with large C_m values (Eq. 50). To keep the noise as small as

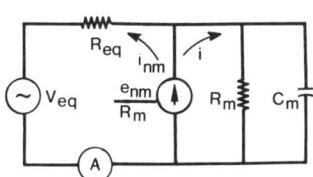

Fig. 21. Thermal noise source of membrane represented by current source equal to $e_{n,m}/R_m$ in parallel with noiseless resistor of value R_m. Current divides as shown. Current through R_{eq} flows through voltage source (V_{eq}) and ammeter. Current (i) through membrane flows wholly within membrane and is not measured.

possible, the experimenter should endeavor to use voltage-recording microelectrodes with the lowest feasible resistances. In addition, attention should be given to keeping the capacitance at the input of the voltage-recording headstage small so that not much capacitance neutralization needs to be used because the neutralization circuit introduces extra noise.

Transfer-conductance voltage clamp

A two-electrode voltage clamp can be implemented with a transfer-conductance (G_T) stage instead of a voltage-gain (A_v) stage in the forward path (e.g., ref. 18 and the chapter by Lecar and Smith). A G_T stage puts out a current proportional to the input voltage irrespective of the output impedance. The circuit and transfer-function diagrams are shown in Figure 22. The transfer function is

$$\frac{V_m(s)}{V_c(s)} = \frac{\alpha_G}{s\tau_G + 1} \tag{53}$$

where

$$\alpha_G = \frac{G_T R_m}{1 + G_T R_m}$$

and

$$\tau_G = \frac{R_m C_m}{1 + G_T R_m}$$

An equivalent circuit can be found; V_{eq} is the value of V_m when $R_m \to \infty$ and $C_m \to 0$. That is

$$V_{eq}(s) = V_c(s) \tag{54}$$

The equivalent source resistance is the value of R_m required to reduce V_m to $V_{eq}/2$. That is

$$R_{eq} = \frac{1}{G_T} \tag{55}$$

Comparing Equations 5 and 53, it can be seen that the two-electrode voltage clamp based on a G_T stage behaves like the two-electrode voltage clamp based on an A_v stage. In both cases the transfer function is the same as that of a first-order low-pass filter. The main difference is that in the G_T-based voltage clamp the DC gain (α_G) and the predominant time constant (τ_G) are independent of R_{e2}.

There are few advantages in using a G_T-based voltage clamp instead of an A_v-based voltage clamp (see the chapter by Lecar and Smith). A

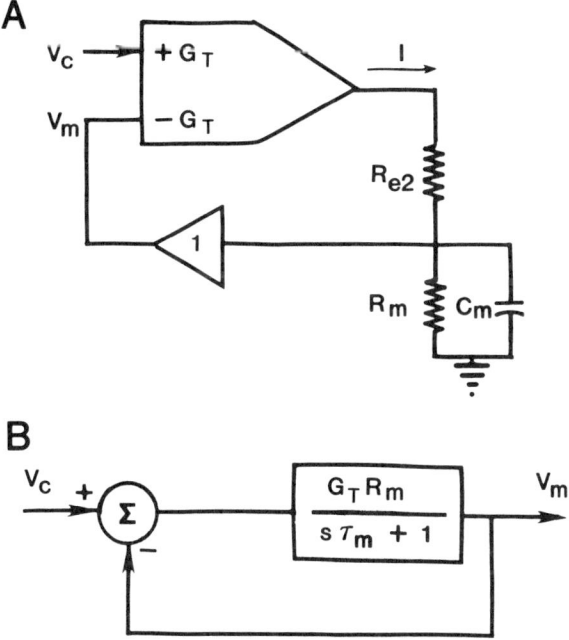

Fig. 22. Transfer-conductance (G_T)-based voltage clamp. *A*: G_T amplifier puts out current equal to $G_T(V_c - V_m)$ independent of value of R_{e2}. Circuit is otherwise similar to Fig. 1. *B*: transfer-function diagram of *A*.

minor practical disadvantage of a G_T-based system is that a larger gain-control range is needed in a general-purpose G_T-based voltage-clamp circuit to enable it to cope with the widely different R_m values encountered in different preparations. In an A_v-based voltage clamp the value of K depends on the relative values of R_{e2} and R_m (Eq. 2) and is independent of the absolute value of R_m. Thus for a particular ratio of R_{e2} to R_m the low-frequency open-loop gain (μK) can be set to the desired value by a given gain-control range in μ. In a G_T-based voltage clamp the low-frequency open-loop gain is $G_T R_m$ (Fig. 22B). Thus, a given gain-control range in G_T will produce an open-loop gain that depends on the absolute value of R_m. A further disadvantage is that shield capacitance (C_{sh}) at the output of the G_T stage alters the performance because some of the controlled output current is diverted to charging C_{sh}. To achieve the ideal response, it is necessary to compensate for the effects of C_{sh}.

The main reason for presenting the equations of the G_T-based two-electrode voltage clamp is to provide a basis of comparison for the single-electrode voltage clamp described by Finkel and Redman. For this purpose it is convenient to calculate the steady-state error (ϵ_1). For a steady-state clamp consisting of V_{eq} in series with R_{eq}, R_m, and R_s (C_m has no effect in steady state), the error is

$$\epsilon_1 = \frac{V_{eq}(R_{eq} + R_s)}{R_{eq} + R_s + R_m} \qquad (56)$$

$$= \frac{V_c(1 + G_T R_s)}{1 + G_T R_s + G_T R_m}$$

It is shown in the chapter by Finkel and Redman that this equation has a direct equivalent in single-electrode voltage clamps.

Discussion

In this paper we present theoretical analyses of a number of different aspects of voltage clamping with two microelectrodes. All analyses use a parallel RC cell model. Consideration of techniques to cope with more complicated membranes is discussed elsewhere (see the chapter by Lecar and Smith). By using the powerful Laplace transform method, several realistic combinations of parameters were studied. Some of the conclusions are listed here.

1. The ideal voltage clamp consists of a parallel RC cell model and infinite-bandwidth recording and amplifying circuits. The fidelity and dynamic response can be made arbitrarily good (i.e., $\alpha \to 1$, $t_r \to 0$) by increasing the clamp gain. Stability is assured at all gains.

2. The voltage-clamp circuit parameters should be optimized to give the best step response of the membrane potential to a square command input. This procedure simultaneously yields the optimum setting for measuring the membrane current responses to changes in conductance or for determining reversal potentials. The only proviso is that the conductance changes must be in parallel with the whole of the membrane under clamp.

3. If series resistance is present it will destroy the equivalence mentioned in the previous paragraph. Techniques to reduce or compensate for the series resistance are presented.

4. A pole (phase lag) can be introduced into the forward amplifying pathway to achieve a critically damped response. At a given gain, the critically damped response is slightly faster than the first-order response without the pole. However, if the pole is removed, more gain can be used and a faster response can be achieved.

5. A zero (phase lead) can be introduced into the feedback pathway to compensate for the limited bandwidth of the voltage-recording microelectrode. Preferably, capacitance neutralization can be utilized to improve the bandwidth of the voltage-recording microelectrode in the first place.

6. Coupling capacitance is the single most important antagonist of good voltage clamping. It is essential to place a grounded shield between the two microelectrodes to reduce the coupling capacitance to manageable or negligible levels. Small amounts of residual coupling capacitance can be compensated for electronically.

7. Unwanted variations in the bath potential can be removed as an error source by using a third microelectrode to measure the bath potential.

8. At a fixed high gain, the stability of the voltage clamp is not affected by the value of the membrane resistance. Thus a voltage clamp established at rest remains stable during membrane conductance changes even if they are large.

9. The most significant source of clamp current noise occurs at high frequencies and is caused by the thermal noise of the voltage-recording microelectrode. The resistance of this microelectrode should be as low as possible to minimize the thermal noise.

Of all these considerations, the two most important requirements for good voltage clamping with two microelectrodes are *1*) a wide-bandwidth voltage-recording microelectrode and *2*) extensive shielding between the two microelectrodes. The other aspects are also highly important, but the degree of their importance may vary depending on the preparation and the experimental arrangement.

We are grateful to Dr. Stephen J. Redman for many helpful discussions and suggestions made during the planning of this paper. We also thank Dr. T. G. Smith and Dr. G. Lamb for constructively commenting on the manuscript.

REFERENCES

1. Adrian, R. H., and M. W. Marshall. Sodium currents in mammalian muscle. *J. Physiol. London* 268: 223–250, 1977.
2. Armstrong, C. M., and L. Binstock. The effects of several alcohols on the properties of the squid giant axon. *J. Gen. Physiol.* 48: 265–277, 1964.
3. Bode, H. N. *Network Analysis and Feedback Amplifier Design.* Princeton, NJ: Van Nostrand, 1945.
4. Cherry, E. M., and D. E. Hooper. *Amplifying Devices and Low-Pass Amplifier Design.* New York: Wiley, 1968.
5. Clark, R. N. *Introduction to Automatic Control Systems.* New York: Wiley, 1962.
6. Desoer, C. A., and E. S. Kuh. *Basic Circuit Theory.* New York: McGraw-Hill, 1969.
7. DiStefano, J. J., III, A. R. Stubberud, and I. J. Williams. *Schaum's Outline of Theory and Problems of Feedback and Control Systems.* New York: McGraw-Hill, 1976.
8. Fishman, H. M., D. J. M. Poussart, and L. E. Moore. Noise measurements in squid axon membrane. *J. Membr. Biol.* 24: 281–304, 1975.
9. Frank, K., M. G. F. Fuortes, and P. G. Nelson. Voltage clamp of motoneuron soma. *Science* 130: 38–39, 1959.
10. Geduldig, D., and R. Gruener. Voltage clamp of the *Aplysia* giant neurone: early sodium and calcium currents. *J. Physiol. London* 211: 217–244, 1970.
11. Hodgkin, A. L., A. F. Huxley, and B. Katz. Measurement of current-voltage relation in the membrane of the giant axon of *Loligo*. *J. Physiol. London* 116: 424–448, 1952.

12. Katz, G. M., and T. L. Schwartz. Temporal control of voltage-clamp membranes: an examination of principles. *J. Membr. Biol.* 17: 275–291, 1974.
13. Lecar, H., and F. Sachs. Membrane noise analysis. In: *Excitable Cells in Tissue Culture*, edited by P. G. Nelson and M. Lieberman. New York: Plenum, 1981, p. 137–172.
14. Meech, R. W., and N. B. Standen. Potassium activation in *Helix aspersa* neurones under voltage clamp: a component mediated by calcium influx. *J. Physiol. London* 249: 211–239, 1975.
15. Moore, J. W., M. Hines, and E. M. Harris. Compensation for resistance in series with excitable membranes. *Biophys. J.* In press.
16. Nyquist, H. Thermal agitation of electric charge in conductors. *Physiol. Rev.* 32: 110–113, 1928.
17. Palti, Y., and W. J. Adelman, Jr. Measurement of axonal membrane conductances and capacity by means of a varying potential voltage clamp. *J. Membr. Biol.* 1: 431–458, 1969.
18. Sachs, F., and H. Lecar. Acetylcholine-induced current fluctuations in tissue-cultured muscle cells under voltage clamp. *Biophys. J.* 17: 129–143, 1977.
19. Sigworth, F. J. The variance of sodium current fluctuations at the node of Ranvier. *J. Physiol. London* 307: 97–129, 1980.
20. Smith, T. G., J. L. Barker, B. M. Smith, and T. R. Colburn. Voltage clamping with microelectrodes. *J. Neurosci. Methods* 3: 105–128, 1980.
21. Stevens, C. F. Inferences about membrane properties from electrical noise measurements. *Biophys. J.* 12: 1028–1047, 1972.
22. Szabo, G. Electrical characteristics of ion transport in lipid bilayer membranes. *Ann. NY Acad. Sci.* 303: 266–280, 1977.
23. Takeuchi, A., and M. Takeuchi. Active phase of frog's end-plate potential. *J. Neurophysiol.* 22: 395–411, 1959.
24. Valdiosera, R., C. Clausen, and R. S. Eisenberg. Measurement of the impedance of frog skeletal muscle fibers. *Biophys. J.* 14: 295–315, 1974.

FIVE

Optimal Voltage Clamping With Single Microelectrode

Alan S. Finkel
Axon Instruments, Inc., Burlingame, California

Steven J. Redman
Experimental Neurology Unit, The John Curtin School of Medical Research, Australian National University, Canberra, Australia

Principles of Operation: Microelectrode voltage artifact, Steady-state clamp error, Clamp stability, Step response, Steady-state ripple in membrane potential, Aliased noise and noise voltage index, Selection of cycle period, duty cycle, and open-loop transconductance • **Practical Considerations in Single-Electrode Voltage Clamp (SEVC) Design:** Design of high-speed low-noise headstage, Smoothing measured current, Selecting gain of controlled current source, Anti-aliasing and output filters to minimize clamp noise, Practical considerations in selection of clamp sample-and-hold amplifier, Setup procedure, Phase control and gain in clamp amplifier • **Examples of SEVC Use:** Voltage clamping motoneurons in cat spinal cord with shielded microelectrode, Voltage clamping smooth muscle in guinea pig mucosa with high-resistance microelectrode • **Conclusion**

The use of two microelectrodes to voltage clamp various nerve and muscle cells is now a well-established electrophysiological technique. But there are limitations in the extent to which this technique can be applied throughout the nervous system. A two-electrode penetration, combined with appropriate shielding measures to minimize electrical interactions between the electrodes, is usually considered too traumatic for many types of small neurons. Also, when using in vivo preparations, it is unlikely that the neurons can be visualized; penetrating such neurons with two independent electrodes or with a two-electrode combination is a difficult task.

These problems of size and geometry can be largely avoided if voltage clamping is achieved with a single electrode. In principle the bridge technique could be modified so that a voltage proportional to the applied current (i.e., voltage across microelectrode resistance) is continuously subtracted from the measured voltage before being compared with a command voltage. This scheme is difficult to implement because microelectrodes do not behave as linear time-invariant resistances; the tip resistance changes as current is passed through it and as it comes in contact with intracellular organelles or the cell membrane.

Another approach is to use a time-sharing scheme in which the electrode is switched from a voltage-recording mode to a current-passing mode. This approach was first developed by Brennecke and Lindemann (1, 2) and was subsequently applied to microelectrodes by Wilson and Goldner (8). This technique, which is now referred to as a discontinuous or switching single-electrode voltage clamp (SEVC), is superior to a two-electrode voltage clamp (TEVC) in several ways. Less cell injury occurs on penetration, and it offers a more practical approach to voltage clamping cells that cannot be directly visualized. Also, because there is no electrode current flowing at the time the cell membrane potential is recorded, only the cell membrane is clamped and errors from any series resistance are avoided. Clamp instability caused by capacitive coupling between two microelectrodes does not occur. However, the SEVC is inferior to the TEVC in two important performance measures. *1*) It takes longer to reestablish steady-state conditions after a sudden change in the command potential or in the membrane resistance. *2*) More noise is present in the current and voltage records.

To utilize the advantages of an SEVC it is important that the inherent limitations of this clamp technique be reduced to their lower limits. This chapter examines the theory of operation of the SEVC and discusses circuit design and operating procedures to maximize the dynamic response of the clamp while minimizing noise.

Principles of Operation

A block diagram and a timing diagram that illustrate the principles of operation of the techniques of Wilson and Goldner (8) are shown in Figures 1 and 2. A single microelectrode is used to penetrate the cell, and the recorded voltage ($V_m + V_e$) is buffered by a high-speed amplifier (Fig. 1, amplifier A1), where V_m is the deviation of the membrane potential from the resting potential and V_e is the voltage developed on the microelectrode resistance and capacitance by the current I_o. At the beginning of the timing diagram (Fig. 2), V_e is almost zero. The sample-and-hold device (SH1) samples $V_e + V_m$ at the time indicated by the arrow in Figure 2, and the sampled voltage (V_{ms}) is held for a complete cycle.

The sampled voltage is compared with the command voltage (V_c) in the differential amplifier (Fig. 1, amplifier A2), and the sample of membrane potential is more negative than required (Fig. 2). The semiconductor switch S1 is connected to the current-passing position and the output of amplifier A2 is applied to a controlled current source (CCS), causing

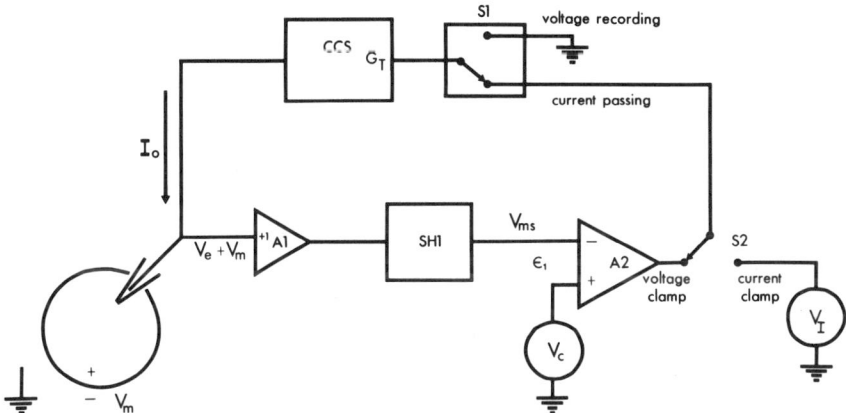

Fig. 1. Block diagram of single-electrode voltage clamp.

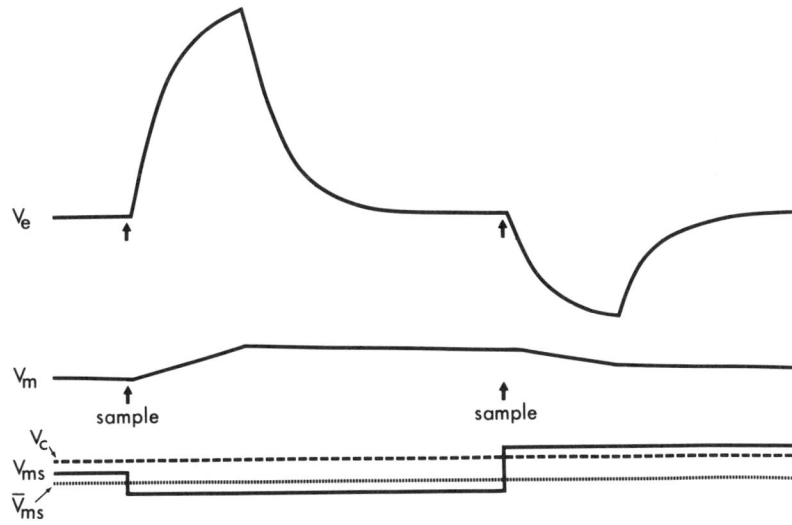

Fig. 2. Switch S1 is closed during current passing and grounded during voltage recording. Microelectrode voltage (V_e) is shown charging exponentially during current-passing period and discharging exponentially during voltage-recording period. Sum of membrane potential (V_m) and V_e is sampled at times indicated by *arrows*. Output (V_{ms}) of sample-and-hold device (SH1) only changes at time of each sample. Comparing V_{ms} and command voltage (V_c) in amplifier A2 determines magnitude and sign of next current pulse. $\bar{V}_{ms}$, average value of V_{ms}.

it to generate a depolarizing current that is applied to the cell via the microelectrode. This current is directly proportional to the CCS input voltage, regardless of the microelectrode resistance. During the current-passing period (T_i) the square pulse of current into the microelectrode causes V_e to rise at a rate determined by the electrode resistance, the capacitance of the wall of the microelectrode to the surrounding tissue or solution, and the input capacitance of amplifier A1. The current pulse would normally last several electrode time constants, but it would be very much shorter than the time constant of the neuron. For this reason a linear change in membrane potential occurs during the current pulse (Fig. 2, V_m). For the present it is assumed that V_e can be described by a simple exponential charging and discharging curve, although for a number of reasons the microelectrode response is nonexponential (see *Antialiasing and output filters to minimize clamp noise*, p. 110).

At the end of the current-passing period, switch S1 changes to the voltage-recording position. In this position the input to the CCS is zero, thus its output is zero. The microelectrode potential decays toward zero while the cell membrane potential decays toward the resting membrane potential. The decay rates are determined by the respective time constants of the microelectrode and cell membrane, and it is assumed in this diagram that these time constants differ by several orders of magnitude. Before a new voltage sample is taken, sufficient time must be allowed for V_e to decay to within a fraction of a millivolt from zero. Because V_e may have reached several volts at the end of a current pulse, the time allowed for the decay of V_e (T_v) must be many (up to 10) microelectrode time constants. (In Fig. 2, changes in V_m are magnified compared with changes in V_e.) Another sample is taken at the end of the decay period, and the cycle is repeated.

Amplifier A2 (Fig. 1) provides negative feedback to clamp V_{ms} to a value very nearly equal to V_c. Under steady-state conditions, V_{ms} moves in small increments about its average value, $\bar{V}_{ms}$. These changes in V_{ms} are caused by system, electrode, and membrane noise sources. The difference between $\bar{V}_{ms}$ and V_c is the steady-state error of the voltage clamp. This difference arises because to maintain stability (see *Clamp stability*, p. 103) the open-loop transconductance (G_T) is finite. The open-loop transconductance is the product of the transfer conductance of the CCS and the voltage gain of amplifier A2.

The CCS and S1 can be used to implement a discontinuous current clamp (DCC) by switching S2 to the current-clamp position (Fig. 2). In this mode the input to the CCS during the current-passing periods is a current command voltage V_I. The resulting current causes a shift in the membrane potential. Provided that the voltage on the microelectrode decays to almost zero during the interval between current pulses, $\bar{V}_{ms}$ is a reliable measure of the membrane potential.

Successful operation of an SEVC depends on the correct adjustment of those clamp parameters that can be controlled. These are G_T, the switching period (T), and the duty cycle (D; fraction of cycle during which current is passed). The electrical characteristics of the microelectrode are also very important, and to some extent the tip resistance (R_e) and the microelectrode capacitance can be controlled by the experimenter. The operation of the clamp also depends on the resistance (R_m) and capacitance (C_m) of the membrane.

The operation of the clamp can be described by a set of simultaneous difference equations. These equations can be used to derive expressions for the various performance criteria of the SEVC, including the stability conditions, the step response, the steady-state and transient errors in clamp potential, and the peak-to-peak ripple in membrane potential. Figure 3 illustrates aspects of the clamp operation during the nth cycle. This figure and Figures 1 and 2 indicate that

$$V_{ms}(n) = V_m(n) + V_e(n) \tag{1}$$

The error in membrane potential at the end of the nth cycle [$\epsilon(n)$] is given by

$$\epsilon(n) = V_c - V_m(n) \tag{2}$$

Because the sampled potential may include a component of electrode potential (if insufficient time is allowed for decay of electrode potential), the error sensed by the SEVC is defined as $\epsilon_1(n)$ and is given by

$$\epsilon_1(n) = V_c - V_{ms}(n) \tag{3}$$

The clamp circuit operates on this error voltage such that

$$I(n) = G_T \epsilon_1(n - 1) \tag{4}$$

where $I(n)$ is the current pulse during the nth period.

The membrane potential at the end of the nth period [$V_m(n)$] depends on its value at the end of the $(n - 1)$th period and on the effect of any additional current during the nth period. It is assumed throughout this chapter that the clamped neuron consists of an isopotential resistance

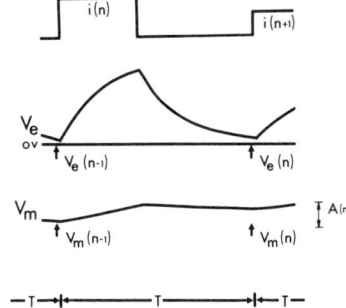

Fig. 3. Current, electrode potential, and membrane potential during nth switching cycle. Values of electrode [$V_e(n)$] and membrane [$V_m(n)$] potentials immediately prior to completion of nth period, when nth sample is taken, are indicated. Zero-voltage reference line for V_m is not shown. $A(n)$, membrane potential ripple during nth period.

(R_m) and capacitance (C_m) in parallel. (No consideration has been given to neurons with significant dendritic processes because the equations become much more complicated in that situation.) With this assumption

$$V_m(n) = V_m(n-1) e^{-\alpha} + I(n)R_m(1 - e^{-D\alpha}) e^{-(1-D)\alpha} \tag{5}$$

where $\alpha = T/R_m C_m$. Similarly, the microelectrode potential at the end of the nth period depends on its value at the end of the $(n-1)$th period and the effect of a current pulse during the nth period. Thus

$$V_e(n) = V_e(n-1) e^{-T_1} + I(n)R_e(1 - e^{-DT_1}) e^{-(1-D)T_1} \tag{6}$$

where $T_1 = T/\tau_e$ and the charging and discharging of the microelectrode capacitance is assumed to be an exponential function with time constant τ_e.

These six equations completely describe the operation of the SEVC and can be used to derive expressions for all the performance measures that have been mentioned. The results are stated simply in the next section.

Microelectrode voltage artifact

If at the end of the sampling period V_e has not decayed to zero, the final value $[V_e(n)]$ is an artifact voltage treated by the clamp circuit as part of the membrane potential. Equation 6 can be used to calculate $V_e(n)$ if all the previous values of $I(n)$ are known. If $I(n)$ reaches a constant value, as would occur under steady-state conditions ($n \to \infty$), Equation 6 can be simplified to give

$$V_e(\infty) = I(\infty) \frac{R_e(1 - e^{-DT_1}) e^{-(1-D)T_1}}{(1 - e^{-T_1})} \tag{7}$$

and $I(\infty) = \bar{I}/D$, where $\bar{I}$ is the average clamp current.

Normally it would be desirable to operate the clamp with an electrode artifact of <1 mV. Figure 4 illustrates the way in which $V_e(\infty)$ depends on D and T_1 for an electrode with R_e = 20 MΩ and an average current of 20 nA.

Equation 7 can be used to calculate the minimum period (t_{min}) that can be achieved consistent with a maximum tolerable value of $V_e(\infty)$. For example, if R_e = 10 MΩ, τ_e = 10 μs, $\bar{I}$ = 20 nA, D = 0.3, and the maximum value of $V_e(\infty)$ is 1 mV, then T_{min} is 92 μs.

Equation 7 can also be used to define an equivalent electrode resistance (R_a), which is used in subsequent expressions

$$R_a = \frac{V_e(\infty)}{I(\infty)}$$

$$= \frac{R_e(1 - e^{-DT_1}) e^{-(1-D)T_1}}{(1 - e^{-T_1})} \tag{8}$$

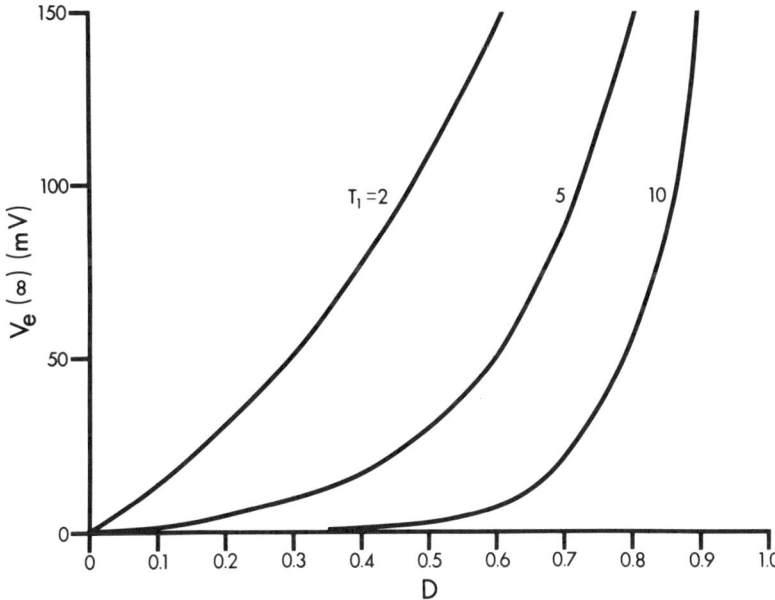

Fig. 4. Equation 7 has been used to calculate microelectrode voltage artifact [$V_e(\infty)$] after many repetitions of same current pulse; i.e., when steady-state conditions have been achieved. It is assumed that *1*) charging and discharging curves are simple exponentials, *2*) DC tip resistance (R_e) is 20 MΩ, and *3*) steady-state current $I(\infty) = 20$ nA and current-passing period $T_i = T/\tau_e$ (where T is cycle period and τ_e is microelectrode time constant). D, duty cycle (T_i/T).

Steady-state clamp error

If R_m and C_m are constant and we assume a stable clamp, it can be shown that

$$\epsilon(\infty) = \frac{V_c(1 + G_T R_a)}{1 + DG_T R_m + G_t R_a}\left[1 + \frac{0.5\ G_T R_m D(1-D)\alpha}{1 + DG_T R_m + G_T R_a} + \ldots\right] \quad (9)$$

where it is assumed that $\alpha \ll 1$ (i.e., switching period is much shorter than membrane time constant). Only the first terms of the expansion are given (brackets) because all other terms have a negligible effect. To minimize the steady-state error, R_a must be negligible; G_T should be made as large as possible, consistent with stability, and D should be made as large as possible. [To maintain as short a cycle period as possible and to reduce the noise in the voltage and current records, the optimum value of D is ~0.3 (unpublished observations).] If an electrode artifact is present, the error can never be reduced toward zero as it can be when no artifact is present. If the electrode artifact becomes very large and if the clamp remains stable, the clamp will eventually cease to clamp the membrane and will clamp R_a instead and the true error voltage will approach V_c.

The SEVC approaches the performance of a TEVC as the duty cycle is lengthened. Letting D → 1 in Equation 9 gives

$$\epsilon(\infty) = \frac{V_c(1 + G_T R_a)}{1 + G_T R_m + G_T R_a} \qquad (10)$$

which is the expression for the steady-state error in a TEVC (see the chapter by Finkel and Gage) if the substitution $R_a = R_s$ is made. Thus it can be seen that the error due to R_a in an SEVC is the same as the error due to R_s in a TEVC.

Figure 5 illustrates how the steady-state error (calculated from Eq. 9) varies with R_m, R_a, and G_T. In this figure, D = 0.3 and α = 0.01. The continuous lines apply when $R_a = 0$ and the broken lines apply when $R_a = 1.74 \times 10^4$ Ω. A small (<1%) steady-state error can be achieved for a given cell membrane resistance by maintaining a negligible electrode artifact and by increasing G_T, consistent with stable operation of the

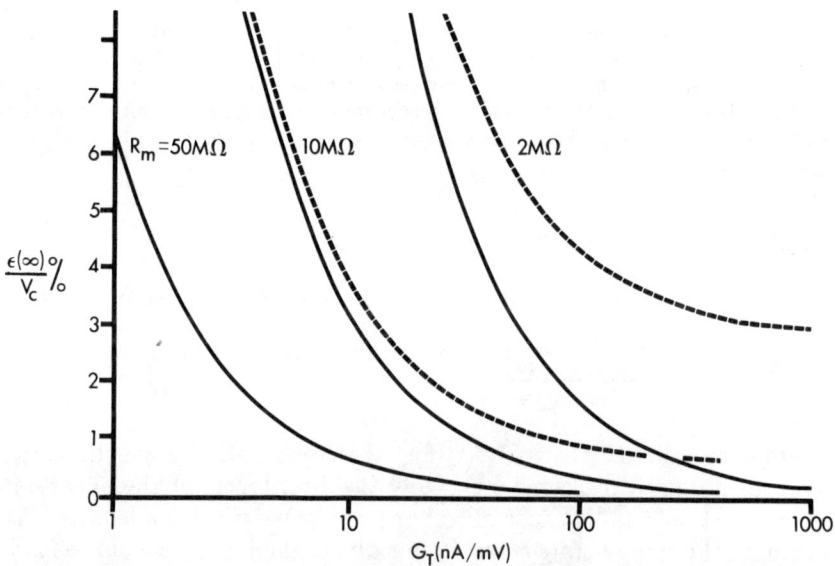

Fig. 5. Steady-state clamp error $\epsilon(\infty)$ [expressed as percentage of command voltage (V_c)] is shown as function of clamp gain (G_T). *Curves* calculated from Eq. 9. *Continuous lines* apply when electrode artifact at time of sample can be neglected: D = 0.3, $\alpha = (T/R_m C_m)$ = 0.01, and R_m = 50, 10, and 2 MΩ, respectively. *Broken lines* apply when $R_a = 1.74 \times 10^4$ Ω, which is obtained for 20-MΩ electrode with T_i = 10 and D = 0.3. For R = 50 MΩ there is negligible difference in *curves* for $R_a = 0$ and $R_a = 1.74 \times 10^4$ Ω. Eq. 9 was derived assuming isopotential cell. D, duty cycle; R_m, membrane resistance; C_m, membrane capacitance; R_a, equivalent electrode resistance; T, cycle period.

clamp. The conditions for stable operation, which impose a limit on G_T, are discussed next.

Clamp stability

The condition for a stable clamp is that

$$0 < \frac{G_T DT}{C_m} < 2 \tag{11}$$

It can be shown that for

$$1 < \frac{G_T DT}{C_m} < 2 \tag{12a}$$

the clamp is underdamped and the error voltage will alternate in sign with each switching cycle before settling to a final value. For

$$0 < \frac{G_T DT}{C_m} < 1 \tag{12b}$$

the clamp is overdamped and will approach its final value monotonically. When

$$\frac{G_T DT}{C_m} = 1 \tag{13}$$

the clamp is critically damped and in theory settles to its final value after only one period. In practice up to two periods are required for the clamp to reach its final value because of asynchrony between the clamp switching times and the instant a step command is applied (see next section).

Thus the maximum open-loop gain consistent with stability is determined by the current-passing period and the membrane capacitance. Figure 6 illustrates the relationship between G_T and C_m (with $D = 0.3$ and various values of T) for critical damping.

Step response

For a step command voltage applied to the clamp, the equations describing the approach of the membrane potential to the new command voltage have been derived by Brennecke and Lindemann (1) and extended by A. S. Finkel and S. J. Redman (unpublished calculations). Figure 7 shows the response of the clamp when it is underdamped, critically damped, and overdamped. The step change in V_c occurred some time

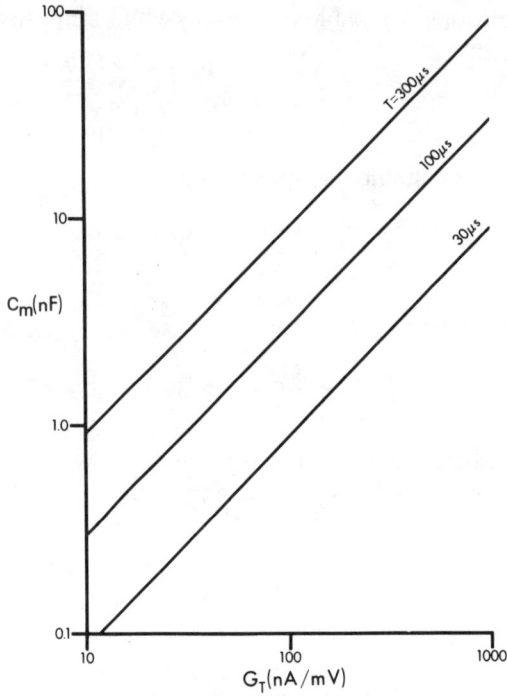

Fig. 6. Curves for different values of cycle interval (T) and for $D = 0.3$ (where D is duty cycle) indicate relationship between total membrane capacitance (C_m) and open-loop transconductance (G_T) for critical damping (Eq. 13).

during the 0th period. It is convenient to normalize the voltage response of the clamp by defining a fractional settling error as

$$\epsilon_s(n) = \frac{\epsilon(n) - \epsilon(\infty)}{V_{ms}(\infty) - V_{ms}(0)} \qquad (14)$$

so that the final settling error is zero, even though there is a final absolute error [$\epsilon(\infty)$]. The expression for $\epsilon_s(n)$ can be shown to be

$$\epsilon_s(n) = \left(1 - \frac{G_T D T}{C_m}\right)^n \qquad (15)$$

The fractional settling error depends on the number of elapsed periods since the step change in V_c and the value of $G_T D T/C_m$. For critical damping, $\epsilon_s(n)$ goes to zero in just one period. Normally the timing of a step perturbation will be asynchronous with the switching times in the SEVC. If V_c undergoes a step change during the 0th period and the system is critically damped, $V_{ms}(1)$ gives the first exact measure of the

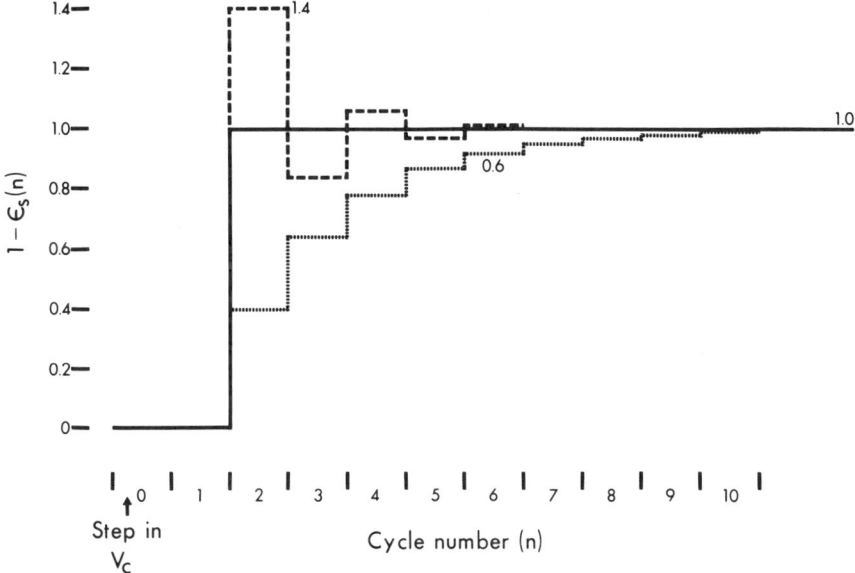

Fig. 7. Step command is applied in 0th period. *Continuous line*, normalized clamp response when clamp is critically damped; *broken line*, response of underdamped clamp when expression in Eq. 12 equals 1.4; *dotted line*, response of overdamped clamp when expression in Eq. 12 equals 0.6. $\epsilon_s(n)$, Normalized error given by Eq. 15.

new steady-state value of V_{ms}. The elapsed time between the onset of the step change in V_c and the first exact measure of V_{ms} will depend on when the step change occurs during the 0th period. Thus the elapsed time between the step change and the first accurate measure of V_{ms} will be between one and two periods in duration. If the system is underdamped or overdamped, more periods will elapse before $V_{ms}(n)$ settles to within a negligible difference from $V_{ms}(\infty)$ (Fig. 7).

The conditions for critical damping and stability are independent of R_m, provided $DT \ll R_m C_m$. This means that a change in R_m will not affect the step response of the voltage clamp when the system is adjusted for critical damping before the change in membrane resistance occurs.

Steady-state ripple in membrane potential

The membrane potential changes during each cycle (see Figs. 2 and 3). Under steady-state conditions the peak-to-peak amplitude of the ripple in the membrane potential will be constant from cycle to cycle, assuming there is no noise in the system. If the membrane potential is to be held constant, the magnitude of this ripple must be very small compared with the command voltage. It can be shown that the steady-state peak-to-peak ripple is

$$A(\infty) = \frac{V_c G_T R_m D(1 - D)\alpha}{1 + G_T R_m D}\left[1 + \frac{0.5\ G_T R_m D(1 - D)\alpha}{1 + G_T R_m D} + \ldots\right] \quad (16)$$

where the electrode artifact is assumed to be negligible ($R_a = 0$). There are constraints on the values that each of the variables in Equation 16 can take if the clamp is to remain stable and operate with a small steady-state error. The latter condition requires that $G_T R_m D \gg 1$ (Eq. 9). Applying this to Equation 16 yields

$$A(\infty) = V_c(1 - D)\alpha[1 + 0.5\alpha(1 - D) + \ldots] \quad (17)$$

The steady-state ripple is reduced as $D \to 1$ and as the clamp is adjusted for overdamped operation. Steady-state ripple also depends on the magnitude of V_c and is zero for a voltage clamp at the resting membrane potential. For example, when $R_m = 10$ MΩ, $C_m = 1$ nF, $D = 0.3$, $T = 100$ μs, and $G_t = 33$ nA/mV, then $A(\infty) = 0.007 V_c$. The presence of ripple means that the average value of V_{ms} achieved by the clamp will be closer to V_c than is indicated by V_{ms}.

Aliased noise and noise voltage index

When a signal is sampled with a frequency f_s ($f_s = T^{-1}$), any components in the signal with frequencies $>f_s/2$ are folded down to the frequency range 0–$f_s/2$. This increases the noise power in this frequency range; this phenomenon is known as aliasing. The frequency $f_s/2$ is known as the Nyquist frequency (f_N). To avoid aliasing, f_N must be $\geq f_i$, where f_i is the bandwidth of the sampled signal.

In an SEVC, f_i is the bandwidth of the signal input to the sample-and-hold device; unless it is limited by an antialiasing filter, f_i is the bandwidth of the microelectrode. While it is possible to arrange for $f_N \geq f_i$, the opposite condition ($f_N \ll f_i$) is necessary to allow $T \gg \tau_e$, which ensures that the electrode voltage at the end of the current-passing period is negligibly small before the next sample is taken.

The noise voltage index (NVI) is a measure of the increase in noise caused by aliasing. It is defined as the ratio of the root-mean-square noise voltage of the sampled signal in the Nyquist bandwidth to the root-mean-square noise voltage of the input signal in the same bandwidth. It can be shown that the minimum value of this ratio is given by

$$NVI_{min} = \left(\frac{T_{min}}{\pi \tau_e}\right)^{0.5}$$

when T_{min} is the minimum period possible, consistent with adequate settling of the microelectrode voltage at the end of each period. Under realistic conditions of SEVC operation, $T_{min} \sim 10\tau_e$ (see Fig. 4) and NVI_{min} is unlikely to be much less than 2.

Selection of cycle period, duty cycle, and open-loop transconductance

The performance of an SEVC depends on the values chosen for T, D, and G_T, and the electrical properties of the microelectrode and cell membrane. The cell membrane properties and the size of the microelectrode tip are constrained by the type of neuron chosen for investigation. The choice of duty cycle is a compromise because small values of D allow smaller values of T, which in turn reduce the aliasing noise and provide for a faster step response; however, small values of D require a greater current-pulse magnitude from the CCS (to maintain given average current). Setting $D = 0.3$ is the best compromise between these conflicting requirements (unpublished calculations).

The electrode time constant can be reduced by shielding techniques, by lowering the bath levels for in vitro preparations, and by using capacitance neutralization (see **Examples of SEVC Use**, p. 116). Figure 4 indicates that the smallest value of T will be $\sim 10\tau_e$ and, because reducing T is paramount in setting up an SEVC, considerable effort must be given to minimizing τ_e.

Once T is reduced to the minimum possible value, G_T should be increased until critical damping is achieved (see Eq. 13 and Fig. 6). When the clamp is critically damped, the fastest step response will be obtained. Any increase in G_T beyond this value will continue to reduce the steady-state error, at the cost of increasing the settling time, as the system becomes underdamped.

Other operational adjustments that contribute to optimum clamp performance (e.g., capacitance neutralization, cut-off frequency of anti-aliasing filter, and phase response of clamp amplifier) are discussed in the next sections.

Practical Considerations in SEVC Design

Design of high-speed low-noise headstage

Possibly the most important part of an SEVC is the headstage. Not only must it be designed and constructed so that the microelectrode bandwidth is large (for rapid settling of microelectrode voltage), but it must achieve this with minimal addition to the unavoidable thermal and 1/f noise of the microelectrode (see the chapter by Finkel and Gage).

Although the headstage performance is of paramount importance in SEVC design, it is also important in other types of voltage clamps.

Because of its general applicability, the considerations involved in head-stage design are discussed in the chapter by Finkel.

Smoothing measured current

The CCS used to inject current during the T_i period may be implemented with the technique discussed in the chapter by Finkel and shown in Figure 8 as a single circuit element. In addition to passing current into the microelectrode, the CCS has an output (V_I) that is a measurement of the output current. During SEVC the CCS output current switches to zero at the sampling rate. The average value of V_I can be obtained by low-pass filtering. Used alone, however, this technique loses a lot of high-frequency signal content. A technique that loses less high-frequency information combines a low-pass filter that has a comparatively high cutoff frequency with a sample-and-hold amplifier (SH2) to sample V_I during each current pulse and store the V_I values until each subsequent pulse (7). To get the average current per cycle, the SH2 output (I_{ms}) must be scaled by D.

Some of the initial current charges the input capacitance of the headstage amplifier (C_{in}) rather than flowing into the microelectrode. This process does not introduce an error in the measurement or control

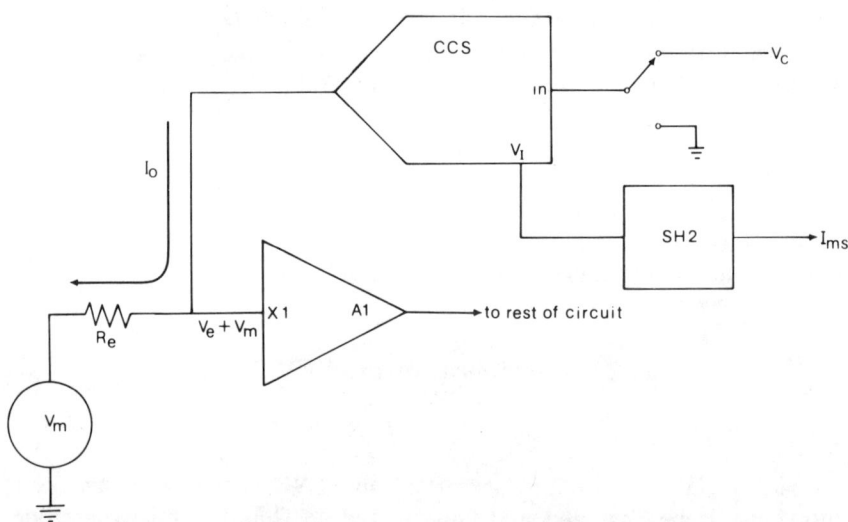

Fig. 8. Circuitry required for current injection and measurement. CCS, controlled current source; SH2, sample-and-hold amplifier. When S1 is in position shown, voltage (V_c) forces CCS to supply current (I_o) equal to V_c/R_o regardless of R_e value (see chapter by Finkel). V_I, output proportional to I_o; V_I is normally proportional to V_c but may be less than that if outputs of A1 or CCS saturate; SH2 samples V_I and holds this value during T_v. Sampled membrane current (I_{ms}) is obtained from V_I by scaling with D.

of I_m because with a stable and accurate voltage clamp the charge delivered to C_{in} during T_i is discharged through R_e during T_v (unpublished calculations). Thus the average value of I_m is equal to the average value of I_o.

Selecting gain of CCS

The current-setting resistor in the CCS (R_o) should be similar to the expected range of R_e values. This choice is made to satisfy conflicting requirements. On the one hand, R_o should be as small as possible because small-value resistors behave more ideally like resistors at high frequencies than do large-value resistors. (The frequency at which the current through the stray capacitance across the resistor becomes comparable to the pure resistive current is inversely proportional to the resistance value.) Also, if $R_o < R_e$, the CCS will not significantly limit the maximum current that can be passed by the microelectrode. On the other hand, $R_o > R_e$ is desirable to minimize the extra noise introduced by the CCS. [This extra noise arises because R_o is part of a positive-feedback circuit that introduces noise in much the same way that a capacitance-neutralization circuit introduces noise (see the chapter by Finkel).]

In addition, R_o should be as large as possible so that the steady-state (DC) error current is minimized. (The DC error current arises because of DC voltage offsets in the CCS that are superimposed across R_o.) In many cases high-resistance cells, which are affected by DC error currents, have slow responses requiring only low-to-moderate sampling rates (1–10 kHz). Cells with fast responses, requiring moderate-to-high sampling rates (5–50 kHz), often have low membrane resistances; therefore they are tolerant of some DC error current. However, in some cases high sampling rates and low DC error currents are necessary and the experimenter may have to be satisfied with a low-gain clamp. To get the best possible high-frequency performance for a given value of R_o, resistors that are physically long and slender should be used. Thick-film resistors made from only a few well-separated spirals of resistance material are much more suitable than metal-film, carbon-film, or wire-wound resistors.

Note that an SEVC could be implemented by using a voltage-source output instead of a current-source output. The theoretical analysis would be different in detail, but equivalent results would be reached. The main reason for using a current-source output is its ease of implementation. *1)* In most experiments the membrane must be current clamped before and after voltage clamping, and a current pulse must be passed down the microelectrode to optimize the capacitance-neutralization circuit. Because a CCS must be included for these purposes it is convenient to use it for voltage clamping also (see the chapter by Lecar and Smith). *2)* The technique of smoothing the measured current, by sampling the pulses and scaling the sampled value, relies on the current of the output stage being constant during T_i. This would not be so for a voltage-source output and the noisier technique of low-pass filtering without sampling would have to be used.

Antialiasing and output filters to minimize clamp noise

The current and voltage noise in an SEVC is inherently worse than in a TEVC because the microelectrode recording circuit must have a wide bandwidth to allow adequate settling and also because the input noise is aliased by the sampling process. The noise arising from the wide bandwidth is minimized by the recommendations outlined in the chapter by Finkel. The noise arising from the sampling process can be reduced in some cases by the addition of an antialiasing filter in front of SH1 to limit the bandwidth of the signal from amplifier A1 (Fig. 9, *top*).

The antialiasing filter is not useful for an ideal microelectrode that settles exponentially after a step in current. In this case both minimum noise and maximum frequency response of the clamp are simultaneously achieved by increasing the sampling frequency to just less than the value that would lead to the onset of instability. However, most real microelectrodes do not settle exponentially because of redistribution of ions in the tip, the distributed nature of the capacitance through the wall of the microelectrode, and other unknown causes. Instead they may settle with both a fast and slow phase as shown in Figure 9, *bottom*. The slow phase limits the maximum sampling rate that can be used. Unfortunately the thermal-noise bandwidth of these microelectrodes seems to be more related to the fast phase than to the slow phase; thus the amount of noise aliased down to the relatively narrow Nyquist bandwidth is large. Considerable reduction of the clamp noise can be achieved by low-pass filtering the output of amplifier A1 to the point at which the fast phase is slowed somewhat but the time for final settling, which is dominated by the slow phase, is largely unaffected. This point is observed by taking the signal indicated as V_{mon} in Figure 9, *top*, to an oscilloscope. It is important to always observe V_{mon} to ensure that the input to SH1 has time to settle properly. Note that V_{mon} is not the same as $V_e + V_m$ unless the filter is set to have negligible effect (i.e., $R \rightarrow 0$).

It might be thought that, if a first-order low-pass filter reduces the noise, a high-order low-pass filter would help even more. However, a high-order *RC* filter or an averager is no better than a first-order filter because, if they are used with the same low-pass cutoff frequency, they have an adverse affect on the stability and dynamic response (personal observations).

Another way to reduce the recorded noise is to include a low-pass filter

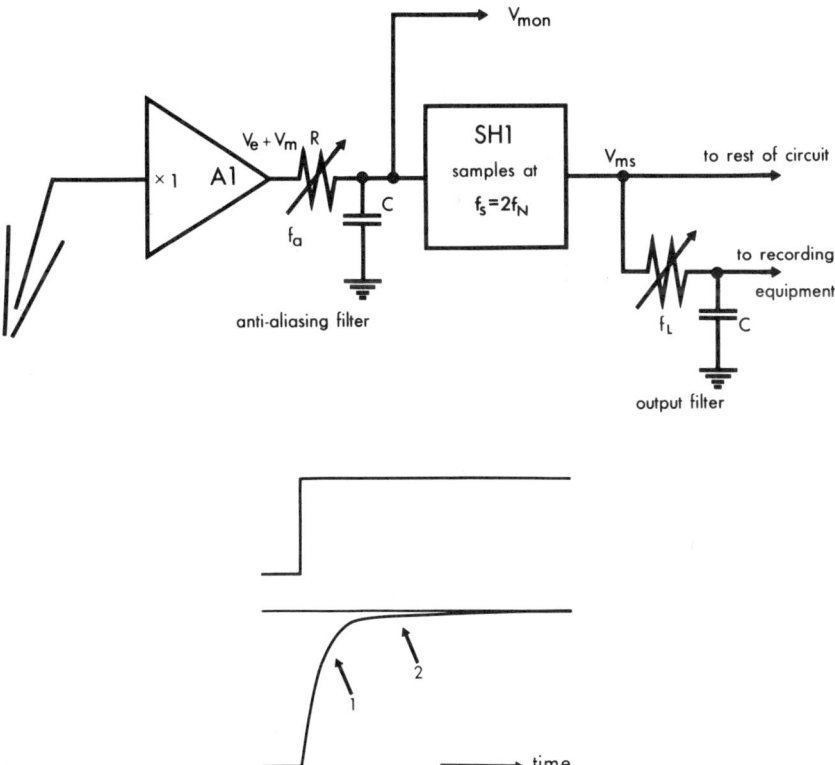

Fig. 9. *Top*: placement of antialiasing filter and output filter. The −3-dB cutoff frequencies $(2\pi RC)^{-1}$ are f_a and f_L, respectively. V_{mon}, input signal to SH1. With microelectrodes that have slow settling phase, antialiasing filter may be used to reduce high-frequency noise content of $V_e + V_m$ without increasing time required for final settling. Settling is checked by observing V_{mon}; f_L is set equal to bandwidth of biological signals of interest. It should be several times less than Nyquist frequency (f_N). *Bottom*: example of nonexponential settling. Voltage response (*lower trace*) of microelectrode after current step (*upper trace*) consists of 2 components. Component 1 is fast phase and component 2 is slow phase, which limits maximum sampling rate.

between the output of SH1 (and SH2) and the recording equipment (Fig. 9, *top*). To be useful the cutoff frequency (f_L) of this low-pass filter must be <f_N, several times less if possible (unpublished observations). If $f_L > f_N$, the bandwidth of the signal to the recording equipment will be determined only by the sampling process. The best way to use the low-pass output filter is to set f_L equal to the bandwidth of the biological signals of interest and to adjust the sampling rate so that f_N is as great as possible.

Even when due consideration is paid to these procedures to minimize noise, the noise in an SEVC is considerably worse than the noise in a TEVC. The noise is greater by at least a factor of two because of aliasing and by a further factor of two because of the high microelectrode bandwidth required. If the noise is too great, signal averaging must be used to reduce the noise without affecting the recording bandwidth.

Practical considerations in selection of clamp sample-and-hold amplifier

A sample-and-hold amplifier is an electronic circuit with two modes of operation, which are selected by a digital control signal. In the sample mode the output closely follows the input. In the hold mode the output is maintained at the value it had at the moment the digital control signal switched from sample to hold. Practical sample-and-hold amplifiers have several relevant limitations. *1)* Droop: during the hold mode the output drifts steadily away from the initial value because of bias currents flowing into the capacitor used to hold the signal; droop can range from much less than one to a few hundred millivolts per millisecond. *2)* Transients: when the modes change, control and switching transients couple into the output; these may range from a few tens of millivolts to several volts and may last from half a microsecond or less to several microseconds. *3)* Acquisition time: this is the time required for a sample to be made; it normally ranges from one to a few hundred microseconds. Generally, sample-and-hold amplifiers with rapid acquisition times achieve this at the expense of their droop and transients characteristics. How brief the acquisition time of SH1 must be for high-speed voltage clamping can be calculated as follows. To obtain low noise and accurate records of clamp currents, the SEVC should be cycled at ~10 times the maximum biological signal frequency. Thus a sampling rate of 50 kHz must be used to obtain good records of currents in a 5-kHz bandwidth, corresponding to a total sample period of 20 μs. For D = 0.3, 6 μs are used for current passing, and 14 μs remain for passive voltage recording. The microelectrode voltage should not be sampled until it has substantially decayed toward a steady value, which usually is achieved only in the last 10% of the voltage-recording period. Thus the maximum allowable acquisition time in this example is 1.4 μs.

To minimize the effects of droop and transients in a sample-and-hold amplifier with such a brief acquisition time, a fixed-gain amplifier should be placed at the output of amplifier A1 [Fig. 9, *top*; (e.g., ref. 7)]. If the fixed gain is G, the importance of the droop and the transients relative to the unamplified input signal is reduced to $1/G$. Values of G ranging from 10 to 50 are practical for sample-and-hold amplifiers that work with ±10-V input signals. To ensure that the amplified input signal stays within the ±10-V limits, a DC offset control must be provided to center the microelectrode signal about zero. This input offset control differs from a normal bucking voltage because it is seen by the voltage-clamp circuit as part of the microelectrode signal. Thus the input offset adjustment must only be made before a cell is penetrated.

Setup procedure

Although each experimenter is bound to develop and prefer his or her own procedures, we recommend this useful starting procedure.

Use two oscilloscopes (CRO). On one CRO observe the outputs of SH1 (V_{ms}) and SH2 (I_{ms}) and trigger from the signal source used to produce the command voltage (V_c). On the second CRO observe V_{mon}. Trigger this oscilloscope at the sample rate (f_s).

Set up a repetitive current step in the DCC mode [i.e., switch S2 (see Fig. 1) in current-clamp position; make V_I a square waveform]. Ensure that the antialiasing-filter time constant is set to its minimum value and that the phase control is switched out or set to have no effect (see next section).

Advance the capacitance-neutralization setting until the step at the leading edge of the V_{ms} response to the current pulse is eliminated. This is equivalent to balancing a conventional bridge current-passing circuit. In each cycle period the waveform on the second CRO will be decaying to its final value before the next current-passing period begins (Fig. 10A). If the capacitance-neutralization setting can be advanced further without causing any overshoot in V_{mon} then do so; f_s can then be increased as long as V_{mon} is allowed sufficient time for complete decay. If the capacitance-neutralization setting is advanced too far, the step in the leading edge of the V_{ms} response will reappear, inverted, because V_{mon} will be overshooting. With further advancement of the capacitance neutralization a critically stable setting may be found in which the ringing on the V_{mon} waveform can be sampled at a point equal to V_m (Fig. 10B, arrow),

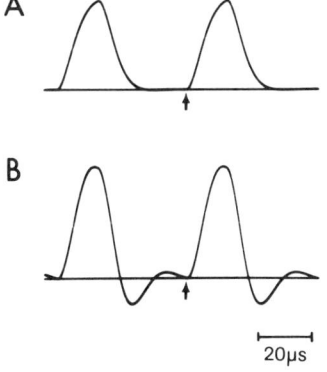

Fig. 10. Value of V_{mon} recorded in cell model at 2 settings of capacitance neutralization during single-electrode current clamp; $R_e = 10$ MΩ. *Horizontal lines* were recorded when current command voltage (V_I) was zero, and transients were recorded when V_I was nonzero. A: capacitance neutralization was optimally adjusted to give rapid settling without overshoot. B: capacitance neutralization was overused so ringing occurred in settling of V_{mon}. At time of sampling (*arrow*), true membrane potential was recorded, but at a critically stable point.

but only for the particular combination of current, sampling rate, and microelectrode resistance used. This critically stable setting cannot be used because even minute changes in any of the parameters will cause instability.

Increase the antialiasing-filter time constant while checking the settling of V_{mon}. With some microelectrodes the noise on the V_{ms} and I_{ms} traces will decrease without compromising the final settling of V_{mon}. If the antialiasing-filter time constant is increased too far, V_{mon} will not settle, and the step in the leading edge of the V_{ms} response will return.

If the step at the leading edge of the V_{ms} response cannot be eliminated at any setting of the capacitance-neutralization control or the antialiasing-filter time constant, decrease f_s. Set the cutoff frequency of the output filter (f_L) to ~10% of f_s.

Set the voltage-clamp gain control to its minimum value, provide a DC command voltage equal to the cell membrane resting potential (see the chapter by Finkel), and switch to SEVC mode. Set up a repetitive step command of ~10 mV, then increase the gain as far as possible without causing instability or overshoot in the step response. Introduce phase lag or lead (see next section) if this will improve the step response of V_{ms} and I_{ms} simultaneously. Monitor V_{mon} throughout the experiment so that changes in the microelectrode settling can be observed and corrected. These changes in settling are normally caused by changes in tip resistance after periods of prolonged current passing.

An example of a correctly established SEVC in a cell model is shown in Figure 11. A second electrode was used in the model to monitor the actual membrane potential, and it can be seen that the actual membrane potential (Fig. 11C) is virtually the same as the sampled membrane potential (Fig. 11A). This necessary equivalence is achieved because V_{mon} (Fig. 11D) settles completely at the end of each cycle.

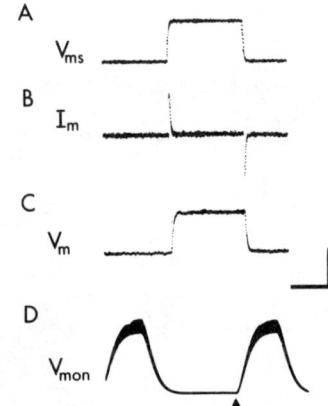

Fig. 11. Example of correctly set up single-electrode voltage clamp in cell model. V_c = 10 mV, 4 ms; R_m = 10 MΩ; C_m = 1 nF; R_e = 10 MΩ; sampling frequency was 15 kHz; and G_T = 40 nA/mV. Calibrations are: A, 10 mV, 2 ms; B, 25 nA, 2 ms; C, 10 mV, 2 ms; D, arbitrary, 20 μs. Records were not averaged. Sampled membrane potential (V_{ms}) and sampled membrane current (I_{ms}) settle in 200 μs (3 periods). Actual membrane potential (V_m) recorded by second nonsampling electrode is similar to V_{ms}. D consists of several superimposed records of V_{mon}. Scatter during current-passing period is from noise sampled by SH1; V_{mon} settles completely at end of each cycle.

Phase control and gain in clamp amplifier

Amplifier A2 (see Fig. 1) is used to provide the negative feedback and amplification required for voltage clamping. It can also be used to modify the phase characteristics of the voltage-clamp circuit. An example of a circuit topology that can be used to vary the phase from lead to lag with a single potentiometer is shown in the chapter by Finkel. The phase is adjustable because as in TEVCs the optimum phase characteristics of the amplifying circuit depend on the frequency and phase response of the membrane being clamped. (No phase control is required to clamp an RC cell membrane.) A special caution must be sounded when phase lag or lead is used in an SEVC. It is possible to find a combination of the capacitance-neutralization setting and phase lag or lead that yields a seemingly fast step response in V_{ms}, whereas in fact the step response in the cell is much slower. This discrepancy between the true intracellular potential (V_m) and the sampled potential (V_{ms}) occurs because the misleading combination requires that the capacitance neutralization be underutilized so that there is a large decay artifact at the end of each voltage-recording period. In effect the SEVC clamps the microelectrode artifact resistance instead of the cell membrane, and the phase control is used to impart stability.

Fortunately there are three simple tests that detect this false condition. *1)* It will be apparent that V_{mon} is not decaying to a horizontal base line at the end of each period (Fig. 12A). *2)* The duration of the I_{ms} response (Fig. 12C) will be substantially longer than the duration of the V_{ms} response (Fig. 12B), whereas in a correctly set up isopotential clamp the duration of I_{ms} and V_{ms} will be similar (see Fig. 11). *3)* The rise time of the V_{ms} step response will depend on the sampling rate.

During a false clamp, I_{ms} remains a true measure of the membrane

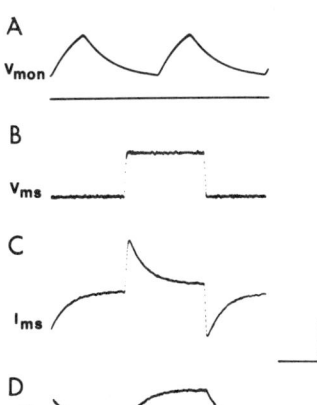

Fig. 12. Criteria for detecting false clamp of microelectrode artifact arising from incorrect adjustment of capacitance-neutralization and phase-shift controls. Records are from cell model used in Fig. 11. Sampling frequency was 20 kHz. Calibrations are: A, 10 mV, 20 μs; B, 10 mV, 2 ms; C, 2 nA, 2 ms; D, 10 mV, 2 ms. A: V_{mon} does not decay to horizontal base line. Duration of I_{ms} response (C) is longer than duration of V_{ms} response (B). Duration of response (D) of actual membrane potential (V_m; recorded by second nonsampling electrode) is similar to duration of I_{ms} response because R_a and R_m are in series.

current because R_a and the cell membrane are in series; therefore the current measurement circuit measures the current through them both. Thus the duration of the I_{ms} response is similar to the duration of the actual membrane potential (Fig. 12D).

This false clamp only occurs if the capacitance-neutralization setting is altered after the phase is changed from its usual flat position. Therefore as a general rule a change in the capacitance-neutralization setting should be avoided once phase changes in the clamp amplifier have been introduced. If the microelectrode characteristics change, necessitating an alteration in the capacitance-neutralization setting, V_{mon} should be carefully watched to make sure that it decays to a horizontal base line at the end of each period.

The open-loop gain of an SEVC is expressed in amps/volt (or for practical convenience in nA/mV), whereas in a TEVC the gain is usually expressed as the open-loop gain of the clamp amplifier in volts/volt. The two are difficult to relate because the interconversion depends on the values of R_e and R_m; i.e., the open-loop gain in volts/volt of an SEVC is 10^6 $(R_e + R_m)$ times the gain in nA/mV. More important than either of these expressions for the open-loop amplifier gain is the closed-loop gain, which is the gain of the system with the attenuation from the cell membrane resistance included. The steady-state closed-loop gain can be measured as follows. When the open-loop gain of a clamp is finite, V_{ms} does not quite follow V_c. The fractional steady-state error is

$$\epsilon_f = [V_c - V_{ms}(\infty)]/V_c$$

and the steady-state closed-loop gain is

$$G_{CL} = \epsilon_f^{-1} - 1$$

Generally G_{CL} should be 10 or more. Values of 100 or more are desirable if quantitative rather than qualitative analyses of results are to be made.

The open-loop gain (G_T) of an SEVC is nonzero only during the current-passing period. The average open-loop gain $(\bar{G}_T)$ is simply obtained by multiplying by D. Only $\bar{G}_T$ should be quoted when comparing results from different experimental situations.

Examples of SEVC Use

Voltage clamping motoneurons in cat spinal cord with shielded microelectrode

Cat motoneurons have cell somata ~50 μm in diameter and an extensive dendritic tree. Most of the measured input resistance of these neurons is from the electrical load of the dendritic membrane. A rough estimate of the resistance and capacitance of the membrane under clamp control would be R_m - 10 MΩ and C_m = 0.5 nF, for τ_m = 5 ms. Motoneurons were successfully penetrated by microelectrodes with tip diameters of 1.5-2 μm. Using thin-walled glass, these tip diameters corresponded to electrode resistances of 2-4 MΩ. The electrode had to be immersed to a depth of ~3 mm in cerebrospinal fluid and neural tissue to reach the motoneuron columns. Because of this, a driven shield insulated from the surrounding tissue was needed around the microelectrode (see ref. 3 for details of electrode fabrication technique). An electrode time constant of 3 μs was usually achieved. This allowed a cycling period of 30 μs. Such an electrode allowed two important conditions to be satisfied: $T \gg \tau_e$ and $T \ll \tau_m$. The duty cycle was fixed at 0.3, and G_T was increased until the step response just started to ring and was then decreased slightly; the capacitance neutralization was previously set to give an optimal monotonic decay of the electrode voltage. On the basis of these figures, and assuming an isopotential soma with no dendrites, critical damping requires a G_T of 50 nA/mV (see Fig. 6); this should have resulted in a steady-state error of ~1% (see Fig. 5). In fact the maximum G_T obtained was 12-15 nA/mV, the steady-state error was in the range of 2%-5%, the average time for a voltage step to reach 90% of its final value was three periods, and the final 10% took even longer to achieve. The steady-state ripple could not be observed in individual responses and was averaged out during repeated responses. The actual higher steady-state error, longer step-response time, and lower maximum value of G_T compared with theoretical predictions emphasize that the theory developed in this chapter applies to clamping an isopotential neuron, whereas for a motoneuron most of the membrane is not isopotential with the soma region.

The soma region of motoneurons was voltage clamped to measure the current evoked at group Ia synapses on these neurons (4). The time course of the excitatory postsynaptic potential (EPSP) evoked by impulses in single group Ia axons was used to determine if the synaptic connection with the motoneuron was on the soma and/or the juxtasomatic region of the dendrites. If it was, the neuron was clamped and the corresponding excitatory synaptic current was recorded. The results of such an experiment are illustrated in Figure 13 in which the *upper record* is the unclamped EPSP, the *middle record* is the membrane potential during voltage clamp, and the *lower record* is the excitatory synaptic current. The synaptic current peaked in 200 μs, decayed with a time constant of 0.33 ms, and had a peak value of 1.3 nA. The dynamic response of the clamp was inadequate to clamp the rising phase of the EPSP. It has been calculated that, if the EPSP had been well clamped, the time to peak would have been ~150 μs and the peak current would

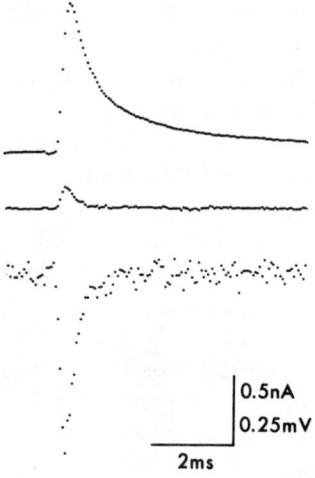

Fig. 13. *Top*: averaged unclamped excitatory postsynaptic potential (EPSP) evoked in spinal motoneuron by impulses in single group Ia axon. Time course of this EPSP indicated that synaptic connections at which it was generated were somatic or juxtasomatic. *Middle*: averaged membrane potential during clamp. *Bottom*: average of clamp current. Each digitizing interval is 62.5 μs. Two thousand records averaged in *middle* and *bottom* responses.

have been increased by 10% to 1.4 nA. The decay phase would not have been altered.

Voltage clamping smooth muscle in guinea pig mucosa with high-resistance microelectrode

Excitatory junction potentials (EJPs) can be recorded in the smooth muscles of the arterioles in the mucosa of the guinea pig small intestine. The arterioles consist of a continuous layer of smooth muscle cells that is only one cell thick. Each cell is 4–6 μm in diameter. The cells are electrically connected to each other and the arteriole has the properties of a one-dimensional cable. The synapses are diffusely located throughout the length of the arterioles.

To interpret currents recorded during voltage clamp, the cell membrane must be isopotential in all the regions containing synapses. Isopotentiality is not possible in a cable that is voltage clamped at a point source, so the preparation had to be cut to yield isolated segments 60–80 μm in diameter and 250–300 μm in length. The electrical length of these segments was only ~0.2 length constants (6). The input resistance at 37°C was 50–100 MΩ and the membrane time constant was 250–500 ms. The periarterial nerves were stimulated directly. Microelectrodes with tip resistances of 80–120 MΩ were used; they were first dipped in silicon oil to prevent creep of the bathing solution up the outer wall of the microelectrode. The sampling frequency was in the range of 1–3 kHz and D was fixed at 0.3; G_T was in the range of 3–6 nA/mV, which was less than required for critical damping (see Fig. 6). The average time for a voltage step to reach 90% of its final value was 2 ms. The steady-state error was usually <2% (cf. 0.5% expected for critical damping; see Fig.

5). However, the error during the rapidly rising phase of the excitatory junction current (EJC) was worse, reaching 10% or more of the unclamped EJP peak.

Some of the properties of the EJPs have been described (6); the aim of this experiment was to record the currents underlying the EJPs (5). The results are shown in Figure 14. The *upper record* is an unclamped EJP, the *middle record* is the membrane potential during voltage clamp, and the *lower record* is the EJC. As can be seen from the presence of a blip in the clamped membrane potential, the clamp response was inadequate to clamp the rising phase of the EJP. However, a maximum duration for the 20%–80% rise time of the EJC was measured from averaged records to be 3 ms. The decay phase of the EJC was well clamped, and the decay of the current was found to be exponential with a time constant of 40–50 ms.

Conclusion

The SEVC is a powerful tool for electrophysiological measurements of cells that cannot be investigated by a TEVC. Because the SEVC is inherently inferior, a TEVC should be used where possible if low-noise and high-fidelity records are required. In some cases though, where the ultimate in dynamic response and low noise is not required (e.g., measurement of reversal potentials), the convenience of only having to penetrate a cell with one microelectrode may make the SEVC the method of choice, even though a TEVC is applicable. In other cases, considera-

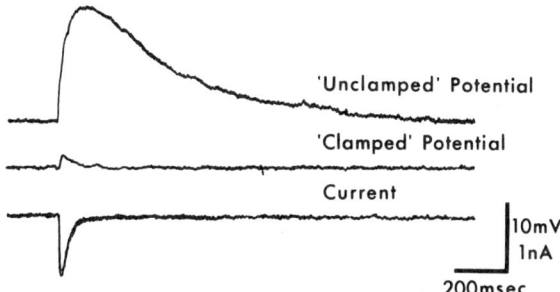

Fig. 14. *Top*: excitatory junction potential was recorded in electrically short segment of arteriole after direct stimulation of periarterial nerves. Arteriolar segment was voltage clamped and nerve stimulated once more. *Middle*: membrane potential during voltage clamp. *Bottom*: clamp current. Clamp was unable to fully clamp rising phase. Decay phase was well clamped and followed exponential time course. Length of arteriole in this experiment was ~200 μm, electrode tip resistance was ~100 MΩ, cycling rate was 1,000 Hz, input resistance of arteriole was ~100 MΩ, membrane time constant was ~300 ms. Twenty records were averaged for each response.

tions of accessibility and penetrability make the SEVC the only practical means of voltage clamping.

To achieve its potential an SEVC must be made from high-speed well-designed electronic circuitry. In use, proper attention must be given to setting the clamp parameters in an orderly way. Once these have been set properly, the most significant factor controlling the quality of the clamp is the microelectrode. Even more than for a TEVC, the SEVC must be used in conjunction with microelectrodes that exhibit extremely fast settling times. To properly clamp currents with 100-μs rise times, the desirable microelectrode (yet to be invented) would have a time constant of <1 μs. Nevertheless even with today's microelectrodes an estimate of submillisecond rise times can be made if some loss of accuracy is acceptable, and membrane currents with time courses of a millisecond or longer can be reliably measured.

REFERENCES

1. Brennecke, R., and B. Lindemann. Theory of membrane voltage clamp with discontinuous feedback through a pulsed current clamp. *Rev. Sci. Instrum.* 45: 184–188, 1974.
2. Brennecke, R., and B. Lindemann. Design of a fast voltage clamp for biological membranes, using discontinuous feedback. *Rev. Sci. Instrum.* 45: 656–661, 1974.
3. Finkel, A. S., and S. J. Redman. A shielded microelectrode suitable for single-electrode voltage clamping. *J. Neurosci. Methods* 9: 23–29, 1983.
4. Finkel, A. S., and S. J. Redman. The synaptic current evoked in cat spinal motoneurones by impulses in single group Ia axons. *J. Physiol. London* 342: 615–632, 1983.
5. Finkel, A. S., D. F. van Helden, G. D. S. Hirst, and S. J. Redman. Excitatory junction currents in arterioles (Abstract). *Proc. Aust. Physiol. Pharmacol. Soc.* 13: 122, 1982.
5a. Finkel, A. S., and S. Redman. Theory and operation of a single microelectrode voltage clamp. *J. Neurosci. Methods* 11: 101–127, 1984.
6. Hirst, G. D. S., and T. O. Neild. An analysis of excitatory junctional potentials recorded from arterioles. *J. Physiol. London* 280: 87–104, 1978.
7. Merickel, M. Design of a single electrode voltage clamp. *J. Neurosci. Methods* 2: 87–96, 1980.
8. Wilson, W. A., and M. M. Goldner. Voltage clamping with a single microelectrode. *J. Neurobiol.* 6: 411–432, 1975.

SIX

High-Resolution Patch-Clamp Techniques

Anthony Auerbach
Frederick Sachs
*Department of Biophysical Sciences,
State University of New York, Buffalo, New York*

Recording Single-Channel Data: Cell treatments, Pipette construction, Seal formation and patch configuration, Characteristics of patch-clamp data • **Instrumentation:** Noise sources, Frequency response, Changing patch potential, Tracking (autozero) operation, Recording potentials • **Analysis of Single-Channel Data:** Theoretical framework, Data acquisition, Low-pass filtering and event detection, Estimating event durations, Fitting histograms, Estimating event amplitudes, Estimating number of active channels in patch • **Summary**

Ion channels are integral membrane proteins that catalyze the diffusion of ions across cell membranes. For example, a single ion channel may pass 20 pA of current, which is equivalent to a flow of $\sim 10^8$ ions/s. Conformational changes in the channel can turn this flow on and off, thereby producing pulses of current. The patch clamp is a technique that can easily resolve these current pulses, making it the most sensitive assay known for the study of protein conformational changes. With traditional macroscopic techniques the kinetic and permeation properties of ion channels are inferred from measurements of the amplitude and time course of currents arising from ensembles of many thousands of channels. With the patch clamp the transition rates between various conformational states of a channel and the current amplitude of each state can be measured directly for individual channels.

The essential feature of the patch clamp is the isolation of a small patch of cell membrane (perhaps a few square micrometers) within the tip of a glass micropipette. The currents flowing into and out of the pipette across the membrane patch are measured, and under appropriate conditions currents through single channels can be detected. There is wide variation in the characteristics of single-channel currents; however, prototypically they are rectangular pulses with an amplitude of a few picoamperes that persist for times ranging from milliseconds to seconds. Some typical single-channel currents are shown in Figure 1.

In addition to its great sensitivity, the patch clamp has several advantages over macroscopic methods of measuring channel currents. The seal between the cell membrane and the electrode is mechanically stable so that a patch of membrane may be excised with either the cytoplasmic or extracellular face of the membrane facing the interior of the electrode. Thus the composition of the solutions along either face of the membrane can be controlled. In contrast to whole-cell techniques, patch-clamp recording from isolated patches does not require a balance of ionic or osmotic strength between the solutions on either side of the membrane. Because the interior of the pipette is chemically isolated from the bath when the patch is still attached to the cell, drugs may be applied either to the patch of membrane in the pipette or exclusively to the surrounding cell membrane. Because the currents are recorded from a small area of membrane, the patch clamp offers a high degree of spatial resolution so that channel characteristics in different regions of a single cell can be measured. The patch clamp can be used to record voltage or current from small whole cells with time resolutions approaching that of axial-wire voltage clamps. Finally, because of the high impedance of the patch relative to the pipette, single-channel currents are virtually free from errors caused by series resistance, incomplete space clamp, or ion accumulation effects.

We briefly discuss some of the fundamental aspects of patch-clamp electrophysiology including cell preparation, electrode construction, seal formation, instrumentation, and analysis of single-channel data. Detailed treatments of patch-clamp technology can be found in Hamill et al. (12), Fenwick et al. (9), and Sakmann and Neher (19).

Recording Single-Channel Data

The key to high-resolution patch-clamp recording is the formation of a high-resistance seal between the cell membrane and the wall of a current-collecting pipette. In practice, seal resistances of $>10^{11}$ Ω (100 GΩ) have been obtained. These high values indicate that the membrane must be in close apposition to the glass, perhaps within molecular dimensions. The salient interaction in seal formation appears to be between the glass and the membrane lipids (rather than membrane proteins or some other element) because high-resistance seals can be made to pure phospholipid membranes.

Cell treatments

To obtain a tight seal the cell surface must be free of connective tissue and if possible free of basement membrane. Intact tissues can be enzymatically cleaned of connective tissue by any one of a variety of methods. For example, snake neuromuscular junctions have been treated for 2 h at room temperature in 2 mg/ml collagenase (Form TD; Advance Biofactures Corp., Lynwood, NY) followed by 20–40 min in 0.02 mg/ml protease [Type XIV; Sigma Chemical Co., St. Louis, MO; (8)] A more aggressive treatment used for *Helix* neurons consists of treating the preparation with 0.5 mg/ml pronase E for 10 min at 20°C followed by treatment in 1% trypsin for 1–2 h at 37°C (14). Alternative treatments include 10–20 min in 0.2% trypsin for *Aplysia* neurons (20) or 1 h in 5% papain for vertebrate cortical neurons (15). Cells grown in tissue culture and loose cells (e.g., red blood cells and mast cells) can be sealed with no enzymatic treatment.

With most cells it is possible to form seals with resistances >10 GΩ. Some cells (e.g., locust striated muscle) resist forming seals above 1 GΩ. The reason for this difficulty in sealing is unknown, but in some preparations (e.g., frog neuromuscular junction) seal formation may be difficult because of the convoluted microstructure of the cell surface. A general observation is that for a given cell type there is a wide variation in the ability to form tight seals from day to day and from preparation to preparation. This variability probably comes from the cell surface rather than from the electrodes, but no controlling variables have yet been defined.

Pipette construction

Patch-clamp electrodes are similar to standard microelectrodes and are only slightly more difficult to construct. A good patch electrode has a steep taper with a tip diameter of 1 μm and a resistance in the range of 1–10 MΩ when filled with physiological saline.

Flint, borosilicate, and aluminosilicate glass appear to have approximately the same noise and sealing properties when properly prepared. Hematocrit or microcapillary glass with an outside diameter of 1.5 mm is satisfactory. A two-stage horizontal or vertical electrode puller can be used for making electrodes. With the horizontal puller the first stage should draw the glass to 200 μm in diameter and the second stage should be adjusted to give a weak pull so the glass has a chance to cool before being broken. With the vertical puller a calibrated mechanical stop is used to limit the first stage travel to 5 mm. The glass is then recentered in the heater and the second stage pull is performed at a lower temperature. For both stages of pull the armature weight is sufficient for drawing the glass without using the solenoid. Electrodes may be made in batches and stored under cover for one day.

The electrodes may be used directly from the puller, but for more reliable use and for lower noise levels the electrodes should be coated with a hydrophobic layer and then fire-polished. A major source of electrical noise is the saline meniscus that creeps up the electrode exterior. This meniscus can be broken by coating the electrode with Sylgard 184 (Dow Corning, Midland, MI), a clear two-part silicone rubber that has excellent insulating properties. Once the rubber and its catalyst have been combined, aliquots of the mixture can be stored in the freezer for several weeks without degradation. The electrode should be coated for a distance of several millimeters, starting 200 μm from the tip. The Sylgard can be applied under a dissecting microscope with a glass wand or hook. It is important to not coat too close to the tip because there apparently are low-molecular-weight components in the Sylgard that can migrate to the tip and interfere with electrode filling and sealing. Also the effect of the coating on reducing the electrode-to-bath capacitance is negligible compared with other sources of input capacity so that there is little to be gained by attempting to coat very close to the tip. After coating, the Sylgard can be cured by advancing the electrode tip into a heated coil or into the flow of air from a hot-air gun. Other materials for coating electrodes (e.g., waxes and lacquers) are inferior to Sylgard. Moisture seems to penetrate between the hydrophobic coating and the hydrophilic glass so that the noise reduction is transient, lasting only 10–15 min. Making the electrode surface hydrophobic through the use of silinizing agents has not proved useful. These agents interfere with electrode filling and the ability to form seals.

Electrodes can be fire polished to burn off contaminating Sylgard and to smooth and blunt the tip, which reduces accidental penetration of the cell. Drawn and coated electrodes can be fire-polished with a standard compound microscope and a heater consisting of a hairpin of 0.005-inch platinum or nichrome wire. A small bleb of soft glass can be melted onto the tip of the filament to prevent the deposition of evaporated metal onto the electrode. A 40× objective (preferably long working distance) is suitable, and standard lenses can be used without damage if minimal heat is used and if the lens itself is protected by a fragment of coverslip. Because the objectives are usually optically corrected for the cover glass, its use also results in better optical resolution. The electrode is held in a lump of plasticine on a glass slide that is manipulated with the mechanical stage of the microscope. The filament is held in a micromanipulator and is heated with a variable voltage source. A Variac voltage transformer connected to a 6V/5A filament transformer is satisfactory. To limit the fire polishing to the electrode tip, the temperature gradient can be sharpened by directing a mild (inaudible) flow of air against the filament through a hypodermic needle.

To fire-polish, the filament is heated to an orange-red and the electrode tip is moved to within a distance of 10 μm of the filament. Thermal convection reduces optical resolution so that the tip is not always resolved and the melting appears as a slight darkening of the pipette walls at the tip. Note that the filament expands as it heats and should be positioned accordingly.

After polishing, the electrodes are filled by dipping the tip into the desired solution that has been previously filtered through a 0.2-μm filter. The first few hundred micrometers of the electrodes are generally filled by capillarity in ~30 s. Filling can be speeded up by applying suction to the back of the pipette. After the tip is filled, the shank is backfilled with filtered solution using a syringe needle. Air bubbles remaining in the pipette are removed by tapping the electrode or by inserting a glass fiber. Filled electrodes do not store well, so they should only be made as needed.

Seal formation and patch configuration

The electrodes should be clamped in a holder that allows for electrical connection of the pipette interior to the amplifier and for the application of suction, usually by mouth. The holder is made of an insulating material such as methacrylate and the electrode should be held firmly enough so that the application of suction does not move the electrode tip. Electrical contact can be established either by a chlorided silver wire (which can be protected from scratching by covering with a piece of small-diameter perforated tubing) or by a silver–silver chloride pellet (which makes a fluid coupling to the filled electrode). Electrode holders are available from commercial sources (e.g., WPI Inc., New Haven, CT). In either case the fluid level should be kept to a minimum to reduce input capacity and to prevent solution from being drawn into the suction line and causing electrical interference.

As the pipette is lowered into the bath, a slight positive pressure is applied to the pipette to prevent the accumulation of debris at the tip. Once in the bath the potential difference between the pipette and the bath is offset to zero. This adjusted potential serves as a reference for changes in patch potential.

The pipette resistance is measured by applying square waves to the amplifier reference input. The amplitude of the resulting current is inversely proportional to the electrode resistance. Because the electrode and seal resistances may vary from 10^6 to 10^{11} Ω, calibration voltages in the range of 0.5–100 mV are necessary to provide adequate resolution.

To make a seal the pipette is gently pressed against the cell surface while the pipette-to-bath resistance is monitored. As the pipette is gradually advanced, the resistance should increase (pulse amplitude will fall) by two to five times its initial value and then level off. At this time, slight suction is applied to the pipette. In good experiments the pipette resistance will suddenly increase to a value >10 GΩ and the base-line noise will be greatly reduced.

Sometimes the application of suction only increases the pipette resistance to several hundred megohms, which is too small for most single-channel recordings. In these cases it is occasionally possible to increase the seal resistance by the alternate application of suction and pressure, but more commonly no tighter seal can be established. Sometimes the application of suction causes the electrode tip to puncture the cell membrane. Penetration results in a large offset in current, an increase in the low-frequency current noise, and an increase in the capacity transient associated with the resistance testing pulse. The probability of penetration is higher with high-resistance or non-fire-polished electrodes. There is also wide variation in the ability of the cell membrane to withstand suction without rupturing. If a high-resistance seal cannot be formed, it is advisable to change electrodes because used pipettes will rarely form tight seals.

Once the seal has been formed, the membrane patch can be excised with the cytoplasmic surface toward the bath (inside-out patch) or with the extracellular surface toward the bath (outside-out patch). To form an inside-out patch, the pipette is simply withdrawn from the cell. Sometimes the membrane patch will form a small vesicle within the tip of the pipette when exposed to the bath solution. The probability of vesicle formation may be reduced by using a low concentration of Ca^{2+} in the bath. A typical bath solution for making inside-out patches is 140 mM KCl, 10 mM N-2-hydroxyethylpiperazine-N'-2-ethanesulfonic acid (HEPES) (pH 7.4), and 0.1 mM ethylene glycol-bis(β-aminoethylether)-N,N'-tetraacetic acid (EGTA). If vesicle formation does occur, it is sometimes possible to rupture the bath side of the vesicle by bringing the tip of the pipette into the air for several seconds.

To make an outside-out patch or to clamp whole cells, the electrode should be filled with a low Ca^{2+} solution. In most preparations the formation of high-resistance seals is not impaired by the removal of Ca^{2+} from the pipette solution. After the cell-attached patch has been formed, the membrane is ruptured by applying a sharp pulse of suction. Membrane rupture is accompanied by an offset current, a decrease in the pipette-to-bath resistance, and an increase in low-frequency noise. In this configuration the current collecting pipette has access to the interior of the cell and the whole cell may be voltage clamped. To form the outside-out patch, the pipette is gradually (few seconds) withdrawn from the cell until the patch is excised. Sometimes this method results in the formation of an inside-out patch even if there is Ca^{2+} present in the bath. In tissue-cultured chick muscle the probability of forming outside-out patches is enhanced by raising the bath concentration of Ca^{2+} to 10 mM (J. Hidalgo, personal communication). The patch configuration can be checked, if the gating properties of the channel are known, by applying the appropriate voltage or drugs to the patch.

In cell-attached patches the pipette potential (V_p) and the bath potential are known but the membrane potential (V_m) is not under direct experimental control. The values of V_m can be estimated from Equation 5 if the single-channel conductance (g) and the reversal potential (V_r) are known. Values of g can often be obtained within the experiment from the slope of a plot of the unitary current versus V_p. However, V_r must be obtained from separate experiments with excised patches or from macroscopic measurements.

When estimating V_m from V_p, it is assumed that the voltage drop across the patch is large compared with that across the rest of the cell membrane. Ayer et al. (3) have used cell-attached patches in conjunction with whole-cell voltage clamps to separately measure the patch and seal resistances in chick cardiac muscle. When the bath and pipette contained normal physiological saline, the resistance of the patch was >200 GΩ and was independent of voltage. When the pipette contained 140 mM K^+ saline, however, the average patch resistance was 18 GΩ and the average seal resistance was 28 GΩ (8 experiments). Thus, when working with small cells or cells with high specific membrane resistances in which the cell impedance is comparable to that of the patch, the membrane potential across the patch may not accurately follow changes in the pipette potential.

Characteristics of patch-clamp data

The prototypical single-channel current is a rectangular pulse that rises and returns to the base line at a rate limited by the system bandwidth (Fig. 1A; see also Fig. 6, *inset*). The open times of channels are commonly clustered into bursts as shown in Figures 1A and 1C. In records in which channel activity is low, the base line can be reliably defined as the current level between bursts; however, when more than one channel is open as in Figure 1B, the base line may be more difficult to define. With more than one channel of a given type active in the patch, the current level will be an integral multiple of the unitary current (Fig. 1B; see also Fig. 6). If the rates of switching between conducting and nonconducting conformations of a channel are comparable to the system bandwidth, the amplitudes and durations of individual events may be difficult to resolve (Fig. 1C). Multiple conducting states may be visible as variable-amplitude closed times (Fig. 1F) or open times (not illustrated). Fluctuations about the mean open-channel current are greater than those about the closed-channel current (background). This phenomenon is probably caused by small conformational changes of the channel and can be seen in Figure 1B as noise, which increases with the number of open channels.

Commonly more than one type of channel will be active in a given

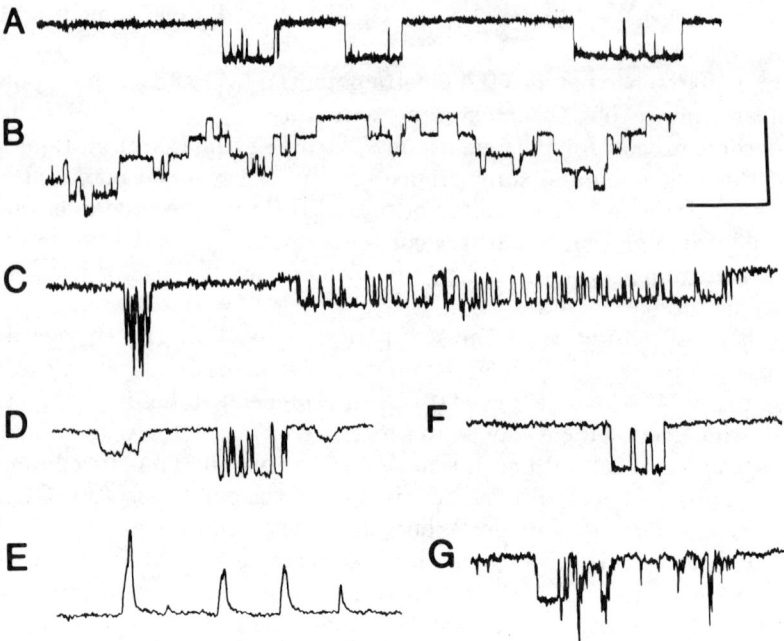

Fig. 1. Single-channel currents from membrane patches attached to chick skeletal muscle cells. Unless otherwise noted, pipette and bath solutions were physiological saline and temperature was 22°C. Inward current is down. *A*: currents from nicotinic acetylcholine receptor channels activated by 200 nM acetylcholine in pipette. Currents are rectangular pulses clustered into bursts. Many gaps within bursts are too brief to be completely resolved and thus appear as spikelike events not reaching the base line. Bandwidth (f_c) was 3.5 kHz and pipette potential (V_p) was +100 mV relative to bath (13 pA, 35 ms). *B*: nicotinic channels activated by 25 nM suberyldicholine in pipette. Amplitude of current at any given time is integral multiple of single current (2.4 pA). Maximum current level is 11.9 pA; thus at least 5 channels were active in patch. Because nicotinic channels are known to carry inward current at cell resting potential, base line was arbitrarily defined as level of least inward current (f_c = 1.25 kHz, 10 pA, 105 ms). *C*: bursts of openings from 2 types of channel in single patch. *Leftmost burst* is from Ca^{+2}-activated K^+-selective channel. Open and closed dwell times in burst are too brief to be resolved, thus true open-channel current cannot easily be determined. (35°C, V_p = +20 mV, pipette contained 140 mM KCl, 10 mM HEPES, and 1 mM EGTA; f_c = 8 kHz, 25 pA, 7 ms). *D*: record showing rounded and square-edged channel currents. Both types of current are from nicotinic channels activated by pentyltrimethylammonium (agonist that can also act as channel blocker). Rounded channels have smaller and more variable amplitudes than do square channels. Difference in size and shape indicates that not all channels in patch have same access resistance to pipette interior (f_c = 1.25 kHz, V_p = +60 mV, 12 pA, 30 ms). *E*: record in which all channel currents are rounded, probably because of formation of vesicle in tip of pipette. Pipette contained same solution as in *C*. Rising edge of channel current is faster than falling edge because during rising edge channel is open and conductance is higher (V_p = −100 mV, f_c = 8 kHz, 18 pA, 40 ms). *F*: nicotinic channel burst showing more than 1 current level (f_c = 2.5 kHz, V_p = +130 mV, 15 pA, 12 ms). *G*: example of seal breakdown. Rectangular pulse to *left* is nicotinic channel current taken from same record as in *F*. Large inward-going spikes are thought to be currents arising from breakdown of membrane and/or seal. These currents are most common when large voltages are applied to pipette. When pipette and bath solutions are symmetrical, these currents reverse polarity when pipette potential equals that of bath. Breakdown can occur in both cell-attached and excised patches.

patch. In some cases these different channel types can be selected on the basis of amplitude (Fig. 1C). The differences can be exaggerated by working near the reversal potential for the interfering channel type. More generally the traditional techniques of ion substitution and pharmacological blocking are appropriate.

Whether the patch is attached to the cell or is excised, several types of distortion affect channel currents. If the membrane patch forms a vesicle within the pipette tip, channel currents may become rounded and exhibit distinct charging time constants (Fig. 1E). Sometimes the channel currents decrease and eventually disappear. If a vesicle is present, this loss of signal may be caused by loss of the voltage gradient and/or depeletion of ions in the vesicle. Additionally the channels may diffuse into the sealing region, or the sealing region may advance to cover the channels. A different pattern is occasionally seen in which some currents appear rounded, whereas others rise squarely out of the base line (Fig. 1D). This cannot be caused by vesicle formation and suggests that some of the channels are in a region that has a high access resistance to the interior of the pipette. One possible explanation is that these rounded currents arise from rim channels located within the sealing region of the patch and thus share current with the pipette and bath. Rim channels are common when seal resistances are <1 GΩ. Alternatively the presence of channels on a highly evaginated or invaginated portion of the cell surface (e.g., T tubule) would also give rise to similarly distorted currents.

Bursts of irregular monopolar noise are sometimes seen, particularly when large driving potentials are applied to the patch (Fig. 1G). The cause of the large spikelike currents is not clear, but we believe that they arise from a breakdown of the membrane or seal.

Generally single-channel data has agreed with macroscopic measurements of channel characteristics; however, there are some situations in which single-channel currents may be distorted by the use of the patch clamp. Because of the need to apply suction and the relatively large area of interaction between the glass and the membrane, channel gating properties may be altered by the patch clamp. For example, the estimates of the density of acetylcholine receptors present in chick muscle are at least 10 times lower from single-channel data than they are from binding and flux measurements (F. Sachs, unpublished observations). Also, at least one type of channel present on chick muscle membranes changes its gating properties in response to membrane stretch and interacts with cytoskeletal components (11). This channel appears to function as a mechanoelectrical transducer; membrane tension could cause subtle changes in the kinetic properties of many types of channels. Finally, differences in channel kinetics have been noted for outside-out, inside-out, and cell-attached patch configurations (10, 22).

Instrumentation

A current-to-voltage converter with high gain is used to record single-channel currents. Simple amplifiers can be built for a few hundred dollars or more elaborate versions can be purchased commercially (Dagan Inc., Minneapolis, MN; List Electronics, represented in the USA by Medical Systems Corp, Great Neck, NY; Yale Instrument Shop, Yale Medical School, New Haven, CT).

The basic headstage is shown in Figure 2. The pipette is connected to the inverting input of A1, an operational amplifier that has a field-effect transistor input stage. For single-channel recording the value of the feedback resistor (R_f) is typically 10 GΩ. For whole-cell recordings the value of R_f may be reduced to 0.1–1 GΩ to avoid amplifier saturation with large currents. The differential amplifier (A2) is a low-voltage noise amplifier with a gain of 10 that subtracts the command voltage (V_c) applied to the noninverting input of A1. The following discussion describes the noise and speed limitations of the patch-clamp amplifier.

Noise sources

The minimum background noise of the amplifier-patch combination is dominated by the thermal (Johnson) noise of all resistances (more precisely, the real part of all impedances) connected to the input of A1. The thermal component of noise is given by

$$I_{RMS} = (4k\text{T}B/R)^{0.5} \tag{1}$$

where I_{RMS} is the equivalent input noise (RMS, root mean square), k is the Boltzmann constant, T is the absolute temperature, B is the noise bandwidth, and R is the parallel combination of R_f and the seal resistance. At room temperature

$$I_{RMS}(\text{pA}) = 0.127[B(\text{kHz})/R(\text{G}\Omega)]^{0.5} \tag{2}$$

Thus low noise can only be obtained with high-value feedback and seal resistances and/or low bandwidths. For a 10-GΩ resistor at room temperature the thermal noise density is $S(f) = 1.6 \times 10^{-30}$ A^2/Hz, corresponding to an RMS noise level of 0.04 pA (0.25 pA peak to peak) in a 1 kHz bandwidth.

Any conductive pathway to ground coupled directly or capacitively (e.g., solution meniscus) will increase the noise level. Thus cleanliness and amplifier layout are important during the assembly of the amplifier. The feedback resistor and A1 should be sonicated in ethanol prior to installation on the circuit board. In humid environments (or where equipment might be subjected to temperature cycles producing condensation) it may be helpful to embed the headstage in Sylgard. To reduce leakage paths to ground the negative input of A1 and the input end of R_f

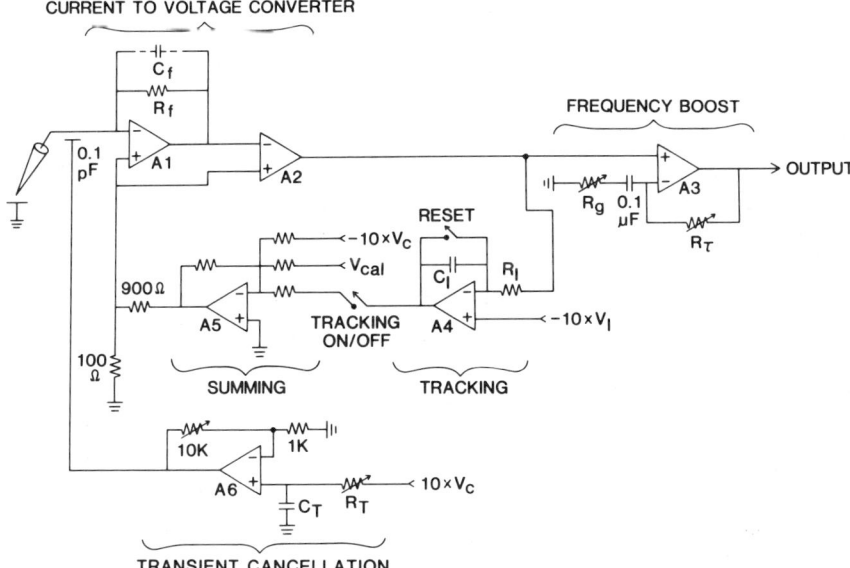

Fig. 2. Patch amplifer. Amplifier A1 determines noise performance of overall amplifier. Either discrete FET amplifiers or Burr-Brown OPA-101 are satisfactory. Open-loop bandwidth of >5 MHz is necessry to avoid having frequency response of first stage depend on input capacity. Amplifier A2 (run at gain of 10) subtracts command potential and should be low-noise type (NE5534, Signetics). Overall frequency response is determined by frequency boost section(s). Amplifier A3 should have low crossover distortion for high-frequency compensation. Value of gain-limiting resistor (R_g) limits maximum frequency of boosting and is typically 330 Ω for highest frequency stage; differentiation resistor (R_r) determines frequency at which boosting begins and is in range of 50 kΩ. Amplifier A4 provides tracking (autozero) mode and current clamp. It should be low input–current device such as LF356 (National). For tracking, integrating resistor (R_I) is ~1 GΩ and C_I is 1 μF. Reset button is used for initial zeroing. For electrometer mode, C_I may be reduced to 100 pF and R_I to 10 kΩ. For constant-current operation, command signals are fed into noninverting input of A4 (V_I). Summing amplifer (A5) adds together command and offset potentials, tracking feedback signal, and electrode resistance measuring pulse (V_{cal}). Summed inputs are applied to A1 through 10:1 voltage divider. Divider helps reduce amplifier noise from earlier stages. Amplifier A5 should be low-noise amplifier (NE5534). Amplifier A6 serves to charge stray capacity of headstage during rapid changes in potential. The 0.1-pF capacitor that couples cancellation signal to input is simply insulated wire wrapped once around input connector. Actual value is not critical; transient cancelling resistor (R_T) may be calibrated to read input capacity. To cancel simple electrode capacity, transient cancelling capacitor (C_T) may be set to zero. To cancel capacitance of whole-cell clamp, additional circuit needs to be added where $R_T C$ is about equal to time constant of cell. Summing amplifier will be needed for adding outputs of multiple cancellation circuits.

should be connected directly to a high-quality Teflon input connector and not wired to the circuit board. The input connectors and electrode holders should also be kept clean by washing them in ethanol or methanol.

The shot noise of the amplifier input current (I_{shot}) produces white noise according to the relationship

$$I_{shot} = (2qB\langle I \rangle)^{0.5} \tag{3}$$

where $\langle I \rangle$ is the input current and q is the electronic charge (1.6×10^{-19} C). An input current of 5 pA produces a white noise contribution equal to that of a 10-GΩ resistor and is generally negligible in wide-band recordings.

A nonadditive noise source comes from current driven through the input capacity by the voltage noise of the amplifier itself (typically 1–5 nV/Hz at 1 kHz) and by noise appearing with V_c. The passive voltage divider attached to the noninverting input of A1 (Fig. 2) reduces the effects of the noise coming through V_c. Because a 1-kΩ resistor has a thermal noise of 4 nV/Hz, which is comparable to the intrinsic voltage noise of the amplifier, low values of resistance are indicated in Figure 2.

Current noise induced by voltage noise seems to be the dominant contribution to the total system noise at higher frequencies. As shown in Figure 3, the high-frequency noise behavior is dominated by a spectral noise density that increases linearly with frequency (f). The linear (rather than f^2) dependence suggests that it arises in a distributed component. A semilogarithmic plot of RMS noise versus bandwidth shows how high-frequency noise dominates the overall noise level (Fig. 4).

A third source of noise is the ever present line-frequency interference. Like other noise sources, this is reduced as the seal resistance increases. To minimize line-frequency interference, it is desirable to use a single-point ground established at the positive input of A1. This point should be used as the only common ground for the amplifier circuits, the bath, and any Faraday shields around the setup.

Low-frequency noise is generated by voices and by building vibration. These vibrations shake the electrode, the bath, and the components in the headstage and induce extraneous currents. The lowest-frequency components may be reduced by mounting the components on a vibration isolation platform.

In practice the noise of the patch recording when the electrode is sealed to a cell membrane is higher than the noise when the electrode is sealed by some other means (e.g., pressing it against bottom of plastic dish). The excess noise density is composed of a low-frequency 1/f component and a high frequency component increasing as f^1-f^2 (Figs. 3 and 4). The 1/f component disappears when the pipette and bath solutions are symmetrical and the potential is zero. This suggests that the low-frequency component is caused by unresolved channel openings and diffusion across the patch or seal. However, even with symmetrical solutions and zero transmembrane potential, there appears to be excess noise above 1 kHz. This noise is probably caused by the distributed

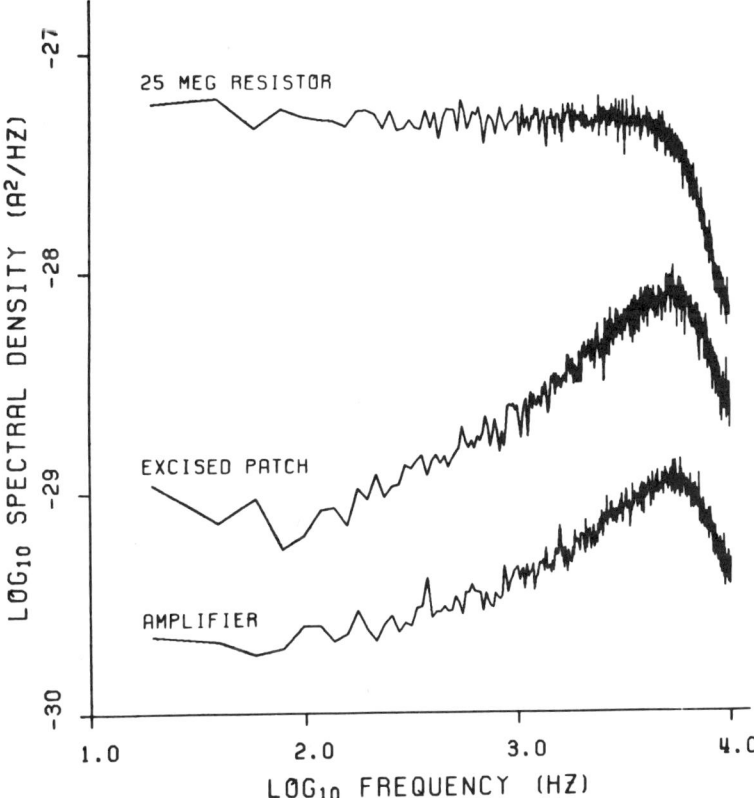

Fig. 3. Spectral density of patch noise under different conditions. For all traces, antialiasing filter (4-pole Bessel) was set at 8 kHz and accounts for rolloff at high frequencies. *Bottom trace* represents patch amplifier with no load at input (amplifer A1 was Burr Brown OPA-101). Noise up to ~500 Hz is white and is caused by noise of 10-GΩ feedback resistor (1.6×10^{-30}) plus equal amount from shot noise of amplifier input current (7 pA). Noise increase above 500 Hz is partly accounted for by product of voltage noise of amplifier and input capacitance, although this mechanism predicts that noise should increase as square of frequency instead of linearly. *Middle trace* shows noise recorded from membrane patch excised in symmetrical solutions with 0 mV applied across patch. Clearly, additive amplifier noise is insignificant. Rough proportionality of patch noise and amplifier noise suggests that they are produced by voltage noise source applied to amplifier input. *Top trace* shows noise produced by 25-MΩ resistor placed across amplifier input. Because resistor noise is white and much larger than intrinsic amplifier noise, resulting spectrum is proportional to squared magnitude of amplifier transfer function and can be used to correct spectra for amplifier and filter response.

properties of the seal impedance, the real part of which is frequency dependent and contributes noise proportional to frequency. The noise at different stages of the recording system is shown in Figure 4. The RMS noise level does not seem dominated by any one stage or component so that dramatic improvements are not expected from improving the noise

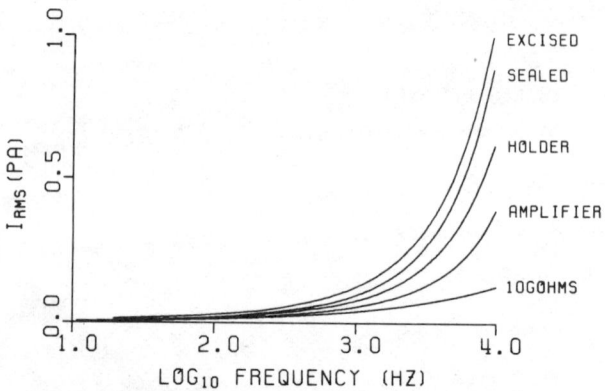

Fig. 4. Root-mean-square (RMS) current noise under different conditions. With exception of curve for 10-GΩ resistor, which was calculated from theory, curves were integrated spectra as in Fig. 3 and were corrected for frequency response of amplifier-filter combination. Curve labeled "10 Gohms" is lowest possible noise level for amplifier using such feedback resistor. Amplifier adds noise to that; the noise increases much faster than expected for resistive load. Curve labeled "holder" shows noise after adding in electrode holder with electrode held just over bath. Immersing pipette in solution for several millimeters and sealing it against bottom of plastic tissue-culture dish (>10^3-GΩ seal) gave "sealed" curve. This additional noise is probably combination of added capacitance from immersing electrode in bath and some dielectric loss in glass of tip. Noise of excised patch in symmetrical solutions with 0 transmembrane potential ["excised" curve is greater than sealed pipette probably because of lossy properties of membrane-to-glass seal (F. Sachs, unpublished calculations)].

performance of the amplifier. Note that the linear ordinate of Figure 4 should not be taken to indicate that the noise from each successive stage is necessarily additive (in sum-squared manner). Current noise induced by voltage noise is multiplicative.

Frequency response

The response time (τ) of the first stage is given by

$$\tau = R_f C_f \qquad (4)$$

where C_f is the unwanted stray capacitance (50 pF) in parallel with R_f (18). A typical value for τ is 0.5 ms. This response time may be adequate for some signals, but it is too slow to study rapid kinetic processes. The bandwidth of the system can be increased simply by adding to the output its derivative (see Fig. 2). This differentiation does not significantly increase the equivalent input noise because the noise contributed by the differentiator is generally much smaller than the noise contributed by the first stage.

Unfortunately, gigohm resistors are not ideal and typically exhibit multiple time constants; thus they require more than one differentiator to properly compensate the frequency response. Resistors will differ in the extent to which they deviate from ideal behavior. Therefore it is important to sample resistors to find one that most closely approaches ideality. We have found that the best commercially available chip (2 × 2 mm) resistors are Type 8-5 from IMS, Inc. (Portsmouth, RI), and the best discrete resistors are Type #K01243 from K&M Electronics (West Springfield, MA).

The frequency response of the amplifier can be tested by air coupling a pure triangle wave to the open input of the amplifier. The triangle will be differentiated by the coupling capacitance to produce a square-wave output. The frequency response can then be adjusted to produce the squarest possible output. It is important to check the settling characteristics with both long (10 ms) and short (0.1 ms) pulses because there may be more than one time constant to compensate. It is simple to obtain fast rise times, but the response to longer pulses may not be accurate. In practice the rise time can be reduced to 15 μs or less with two successive stages of differentiation. If power spectra can be calculated, the amplitude of the transfer function can be simply measured by using as a white-noise source a relatively low value resistor (25 MΩ) connected between the input of A1 and ground (see Fig. 3). Note that this method is not sensitive to phase distortions.

Finally the linearity of the system should be checked by applying different amplitude signals to the input. We have observed nonlinear behavior in the form of a slow creep of the output and in the form of ringing. The ringing seems to be associated with crossover distortion in the differentiating amplifiers and is worst when attempting excessive frequency boosting.

Changing patch potential

Voltage signals may be applied to V_c to control the patch potential (see Fig. 2). The accuracy of the voltage command is given by the ratio of the pipette resistance to the seal resistance. Because seal resistances are commonly 100 times larger than pipette resistances, the DC voltage error is negligible. When seals are not tight or when whole cells are being clamped, there may be substantial errors. The most common limitation to applying voltage is that with low resistances the headstage amplifier saturates and noise becomes excessive.

For high-speed changes in potential, transient compensation is required. As shown in Figure 2, the voltage-clamp speed limitation comes from the current required to charge the input capacity. The input capacity may be 10 pF or more, so that for a feedback resistance of 10 GΩ, the charging current cannot settle much faster than 0.1 s. If faster changes in potential are required (e.g., in studying voltage-dependent channels), the input capacity must be charged from an external source.

Figure 2 shows a typical transient cancellation circuit. This circuit capacitively couples to the input of A1 a signal with the proper amount of charge to raise the input capacity to the required voltage. Because this charge comes from a low-impedance source, A1 is not required to supply the current. In practice, two or three parallel stages may be necessary to satisfactorily compensate the different time constants associated with the pipette capacity and the membrane time constant. With proper compensation, rise times of 20–30 μs are possible when voltage clamping small cells. Series-resistance compensation may be required if the current-resistance (IR) drop through the pipette is significant with respect to the applied potential.

Tracking (autozero) operation

When the patch electrode is first placed in solution, the amplifier may have a voltage gain of 10^4–10^5, depending on the electrode resistance. Small voltage offsets or drift (~100 μV) can cause the amplifier to saturate or wander over the oscilloscope screen. To simplify the initial setup, amplifier A4 serves as an integrator that "bootstraps" the input so that there is no mean output. This autozeroing, or tracking, circuit effectively makes the system AC coupled with a time constant of $R_I C_I / G$, where G is the net gain of A1 and A2, which depends on the electrode resistance and R_f. The integrator is also useful for recording cell potentials.

There are two types of errors in potential application and measurement: those from diffusion potentials and those from amplifier input current. When there is a difference in ion concentrations between the pipette and the bath, some of the offset potential (applied to A5 through V_c; see Fig. 2) used to initially zero the amplifier was used to counter the diffusion potential. The diffusion potential may add to or subtract from the applied potential, depending on the ion gradients, and should be allowed for in precision measurements (9). For a solution of 140 mM NaCl in contact with 140 mM KCl, the diffusion potential is 5 mV, with the Na$^+$ side positive because of the lower mobility of Na$^+$ compared with K$^+$. The diffusion potential may be strongly temperature dependent if strongly associating ions (e.g., fluoride) are present. Because of the unknown selectivity properties of the seal, DC measurements of patch currents are highly uncertain.

Amplifier input current can cause a correctable error in the DC patch current. When the pipette is first placed in solution, the input current flows to ground through the pipette resistance, producing a voltage drop of only a few microvolts (5 pA $\times$ 1 MΩ). When a high-resistance seal is formed, the current must be supplied through the feedback resistor. This error is only significant for DC measurements and can be compensated if the input current and seal resistance are known. If the offset voltage trim of A2 is adjusted so that its output is zero with no input load on A1 and A1 is offset (by biasing command voltage) with the pipette in solution, the DC output will no longer be sensitive to changes in input resistance.

Recording potentials

With whole-cell patch clamps it may be useful to measure the cell potential. Feedback can be used to convert the patch-clamp amplifier to a current clamp. If the output voltage is fed back through V_c (see Fig. 2) so that it reduces the input current to zero or some other constant value, the system behaves as an electrometer capable of constant current injection. In Figure 2 the time constant of the integrator A4 can be reduced so that V_c tracks the input signal to A1. In the voltage-recording mode the amplifier can follow signals with a time resolution limited by the input capacity and the electrode resistance just as in an ordinary electrometer.

Analysis of Single-Channel Data

The first stage of data analysis consists of acquiring, storing, filtering, and detecting events. The next step is to measure the amplitude and duration of individual events. Finally, these data are collated and then fit to appropriate functions that characterize the population. An excellent discussion of these aspects has been given by Colquhoun and Sigworth (7).

Theoretical framework

The measurement of channel current amplitude is the simplest and least ambiguous of single-channel parameters. The magnitude of single-channel currents can be characterized by the ionic specificity of the reversal potential and the channel conductance. The reversal potential is that transmembrane potential at which no current flows through the channel and may be the same as the Nernst potential if only one ion species is permeable. The conductance is then defined as

$$g = I/(V_m - V_r) \qquad (5)$$

where I is the observed channel current. The conductance may vary with ion species, concentration, and membrane potential. Channel conductances typically vary from 1 to 300 pS.

For kinetic analysis the individual channels are usually assumed to undergo transitions between conformational states with time-independent probabilities. The time a channel resides in a given state (dwell time) is an exponentially distributed random variable. The mean duration of a given state is equal to the inverse of the sum of all possible exit rates from that state. For example, consider a channel that can be in either of two states: closed or open. The mean dwell time in the open state would be equal to the inverse of the channel closing rate and the mean dwell time in the closed state would be equal to the inverse of the channel opening rate. To arrive at estimates of these mean durations, many events need to be measured because the standard deviation of an exponential distribution is equal to its mean.

If all states of a channel were distinguishable (i.e., by amplitude), then estimating transition rates from single-channel data would be a relatively simple matter of determining the mean dwell time in each state and measuring the ratios of the transition probabilities between different states. However, most channels can exist in more than one state that cannot be unambiguously identified. The unconditional distribution of dwell times in these nonsingular states is described by a sum of exponentials, the number of which equals the number of states. For example, in the following model of a single channel

$$\text{Closed}_1 \underset{k_{-1}}{\overset{k_{+1}}{\rightleftarrows}} \text{Closed}_2 \underset{k_{-2}}{\overset{k_{+2}}{\rightleftarrows}} \text{Open}$$

where k is the transition rate constant, the open times will be distributed as a single exponential with a mean equal to $1/k_{-2}$, and the closed times will be distributed as two exponentials with a mean equal to $(k_{+1} + k_{-1})/(k_{+2}k_{-2})$. The mean closed time will be longer than the dwell time in either of the closed states because transitions between closed states are undetected. As shown in Figure 1A, channel events may consist of bursts of closely spaced openings separated by short gaps. The presence of bursts shows that the channel can exist in at least two closed states because there are two kinetically distinguishable populations of gaps. In this model the short gaps within bursts would reflect residence of the channel in Closed$_2$. These gaps would have a mean duration equal to $1/(k_{+2} + k_{-1})$. Thorough treatments of the stochastic properties of single channels as Markov processes can be found in Colquhoun and Hawkes (5, 6). The analysis of multiple-channel activity may be possible with the techniques developed by Dionne and Leibowitz (8) and Horn and Lange (13).

The analysis of kinetic data is further complicated if more than one channel is active in the record because the times between observed transitions often cannot be attributed to a single channel. For example, in the record shown in Figure 1B, virtually none of the dwell times in any of the six observable current levels can unambiguously be associated with the residence of an individual channel at that level. However, even when more than one channel is present, some dwell times at a given current level can be associated with a single channel. For example, the bursts of open periods shown in Figure 1A–C and the brief gaps within those bursts can with confidence be attributed to a single channel because the probability of several independent openings separated by such short durations is extremely small.

These considerations of channel kinetics have assumed that the data is stationary; i.e., the statistics in question do not change with time in the record. In practice the kinetics often show drift. Rates may change because of the nature of the channel kinetics, as in the case of Na^+ channels, or perhaps because of changes in the environment of a channel. When the kinetics of a single molecule are being observed, subtle changes must be expected; these changes are averaged out in macroscopic records. The extreme nonstationarity in Na^+ channel kinetics, for example, may be treated with conditional probabilites that explicitly include time (1). With more subtle forms of nonstationarity (e.g., drug desensitization and slow drifts in kinetics) care must be used in the analysis and caution in the conclusions.

Data acquisition

The raw data may be collected in different ways depending on the type of experiment and the facilities available. For continuous random data, as observed with drug-activated channels, the analog tape recorder provides a very efficient means of storing data. Hours of data can be stored on a reel of tape, and several data channels are available to record voice comments as well as synchronization marks. Analog tape recorders are expensive, however, and don't provide a means of analysis.

For data that are not continuous (e.g., Na^+ channel activity induced by voltage steps) the tape recorder is inefficient because it does not function well in short start-stop operation. For this synchronized data a computer or transient recorder provides a more efficient means of data acquisition. If only a single device is to be purchased, there is little question that a computer with disk or tape storage is the best choice.

In some cases the data may be so reproducible and of such high signal-to-noise ratio that simple analog detection circuitry may be used to measure kinetics. This eliminates the need to store the raw data but makes it impossible to analyze the amplitude characteristics of the current or to reexamine the data should some unexpected effect be observed.

The simplest recording device is the strip chart recorder with a pair of calipers for measurement. If the data is sufficiently slow, the recording may be made directly. If the data is arriving too fast, a variable-speed tape recorder may be used to slow down the data. The strip chart may still provide the best global view of a long segment of data.

Computerized analysis methods vary from totally interactive systems in which the user positions cursors to mark features in the data to automated pattern-recognition systems, the choice being made by available equipment and software (7, 17). Regardless of the system used, some features of the data collection and analysis are common to all.

Low-pass filtering and event detection

In the first phase of data analysis the data are band limited with a low-pass filter and digitized so that opening and closing transitions can be detected. The requirements for these different stages of analysis are interrelated so that it is convenient to define the parameters to be used. Let f_c denote the cutoff frequency of the filter. For consistency, f_c refers to the 3-dB point of the filter, the frequency at which the signal power is reduced to half of its zero-frequency value. The time constant of the filter (τ) is equal to $1/(2\pi f_c)$. The rise time is taken as the inverse slope of the response to a step; for a four-pole Bessel filter this is ~$0.3/f_c$. The digitizing rate (f_s) is the inverse of the sampling interval (Δt).

The purpose of low-pass filtering is to limit the noise level so that false positive detections are minimized and channel currents can be clearly visualized. After analog filtering, digital filters may be used to further limit the bandwidth. There is no clearly optimal filter for single-channel records because both the signal and the noise are broad band and the appropriate filter depends on the the type of measurement being made. The main requirement is that the filter order be higher than two so that the high-frequency noise, whose power increases linearly with frequency (see Fig. 3), is adequately reduced.

Inevitably low-pass filtering causes short-duration events to be attenuated. For the same f_c, sharp cutoff filters (e.g., Butterworth filter, flat amplitude) afford a lower noise and a slower rise time than do flat delay filters (e.g., Bessel filter). The lower noise of the sharp cutoff filters means that fewer false detections will occur, but the slower rise times mean that more short-duration events will be missed. Perhaps the main reason for using slower rolloff filters is that their response to a step input does not overshoot significantly, so that the bias they produce in channel amplitude is in a consistent direction.

The cutoff frequency of the filter is determined by the maximum allowable rate of false positive detections. For Gaussian background noise the number of threshold crossings per second is closely approximated by

$$R = f_c/2 \times \exp[-0.5(a/\sigma)^2] \qquad (6)$$

where R is the false positive rate, σ is the RMS variance of the noise, and a is the detection threshold. The allowable value of R depends on the channel transition rate. Because false positive detections cause errors in the observed distribution of event durations, the slowest transition rate of interest should be much greater than R. Assume the detection threshold is set to half the channel amplitude. The mean time between false positive detections (1/R) is 100 ms/kHz of bandwidth when the peak-to-peak noise level (taken as 6σ) is equal to the single-channel amplitude. This value increases to 36 h/kHz when the channel amplitude is twice the peak-to-peak noise. With these extremely low false positive rates, however, the assumption that the noise is Gaussian is no longer correct. The presence of base-line drift, unresolved channel currents, and power-line interference becomes limiting. Also, the current fluctuations of the open channel may be much greater than those of the closed channel; thus false closures would be more likely than false openings. As a rule of thumb, to obtain a tolerable false positive rate the peak-to-peak noise of the base line should be $<2/3$ of the single-channel amplitude.

If the data are to be digitized for computer analysis, they should be sampled at two to five times f_c. The lower limit is the Nyquist frequency and is the minimum sampling frequency that can recover a sine wave of frequency f_c. The higher the sampling rate, the more accurately the digitized data represents the input data. However, above the Nyquist frequency the extra information gained approaches zero because the signal amplitude between sample points can accurately be predicted from knowledge of the filter response. A sampling rate of five times f_c appears to be the highest useful rate for most applications (7). For a fixed amount of storage space, higher sampling rates allow higher precision for individual events, but fewer events can be recorded; this results in less precision in estimating the population parameters. The effect of the sampling rate on the observed distribution of event durations is considered in next section.

The optimal threshold for detection is one at which the number of detected events is maximal and the false positive rate is small. It can be shown that for a noise-free band-limited exponential distribution of durations the maximum number of opening and closing transitions are detected when the threshold is at one-half the single-channel current. In the presence of noise the probability of detection is less sensitive to the precise position of the threshold. Because the half-amplitude threshold has no hysteresis, it may cause false triggering from noise that is present on the rising and falling phases of the pulse. This kind of false triggering is only a problem when the data has been oversampled because with lower sampling rates the transitions are complete within a single sample period.

If an automated-analysis computer program is used, two further considerations in event detection are necessary. *1)* Base lines are not perfectly flat, so an accurate and robust algorithm is needed to define the base line. *2)* Additional criteria should be applied to detected events before a putative event is accepted for further analysis. This validation is necessary because records often contain currents from more than one type of channel (see Fig. 1C) and "noisy" events such as the seal breakdown shown in Figure 1G. Two useful criteria are that *1)* the event amplitudes lie within some window of current centered around the amplitude of the event of interest and *2)* that the RMS deviation of the data from an idealized rectangular-pulse model of the event be less than some critical value. It is important to note that, when there is more than one type of ion channel in a patch, the current amplitude of the channels may be similar. Thus the channels must be distinguished by some other set of criteria besides amplitude (e.g., kinetic or selectivity properties).

Estimating event durations

A simple and rapid method of determining event durations is to measure the time between half-amplitude crossings. Systematic errors in determining the parameters of the population arise because durations of brief events that do cross threshold are underestimated, and events with amplitudes less than half the single current caused by filtering are not measured. Figure 5 shows a plot of the observed duration distribution for an ideal exponential process that has passed through a single-pole band-limiting filter. The event durations are taken to be the time above threshold, and the calculation is done without noise to simplify the results. No events shorter than $0.1/f_c$ are recorded, and events up to $0.3/f_c$ will have measured durations significantly less than the true event duration. For data band limited by an eight-pole Bessel filter, events shorter than $0.2/f_c$ (~30 µs, bandwidth 5 kHz) do not reach half amplitude, and only events longer than $0.25/f_c$ (50 µs, bandwidth 5 kHz) will be measured with >90% accuracy. The addition of noise will broaden the observed distributions and will extend the minimum duration at which the data will match the true distribution. As a rule of thumb, the observed duration distributions should only be fit to exponential functions from times equal to $0.5/f_c$ or longer. When the sampling rate is equal to $2f_c$, this allowance is equal to Δt.

Fitting exponentials to the observed curves can either overestimate or underestimate the population distribution parameters because the observed distributions are not really exponential (see curve 0.1 in Fig. 5B). Note, however, that adequate exponential fits may be obtained to data that are monotonic (e.g., curves 1 and 3 in Fig. 5), but the extracted time constant will be too long by a factor of two or more. With multiple exponential processes the errors in amplitude and duration from band limiting will change the proportion of the fast and slow events.

For the sampling rate, if the observed distribution of durations is to be compared with theoretical models that do not include the effects of the recording system and noise, then sampling at the Nyquist limit ($f_s = 2f_c$)

is satisfactory and, if storage space is limited, provides better precision for estimating the probability distributions than if the data were oversampled. Sampling at the Nyquist limit produces scatter in the measured durations of individual events because the sampling clock is not synchronized with the channels. However, the duration of individual events is of little interest because only population means have testable significance. For exponential distributions it can be shown that the scatter does not bias or significantly alter the errors in the derived parameters.

An error in duration measurement occurs when short events are missed because the period including the missed event appears longer than it actually is. For example, if a short closed period is missed, the apparent open time for the adjacent events is increased. If we assume that both the open and closed periods are exponentially distributed, that closed periods are much shorter than open periods, and that events shorter than some dead time (T_d) are missed, the observed distribution is still exponential, but the observed rate constant is smaller than the true rate constant. The relationship is

$$T_t = T_o \exp[-(T_c/T_d)] \qquad (7)$$

where T_o is the observed open time constant, T_t is the true open time constant, and T_c is the true closed time constant.

Fitting histograms

There are a few points to keep in mind about fitting histograms. *1)* The population distortions discussed above limit the shortest events that can be included in the fit. *2)* The residuals [(data minus theoretical)^2] must be weighted inversely with the data variance (4). For histograms this variance is equal to the number of counts in each bin because bins are filled in a Poisson manner. *3)* Histograms containing empty bins must be reformatted so that each bin to be fit contains at least five counts. The usual equations for nonlinear regression assume that the data variance is approximately Gaussian, and a minimum of five counts per bin has been found to be adequate to give an unbiased estimate of the population mean. Bins should have variable spacing to preserve resolution at short times. To avoid errors caused by finite bin widths, the theoretical function should be the integral of the appropriate probability density taken over each bin. *4)* To have confidence that the fitted parameters can be trusted, error estimates on the parameters of the fit should allow for cross-correlation between parameters. This cross-correlation means that to estimate the time constant of a single exponential distribution to within 10% of its true value with 90% confidence, 500 or more counts are required. For multiple exponentials no simple guidelines exist. To get some idea of the errors in fitting even a double exponential

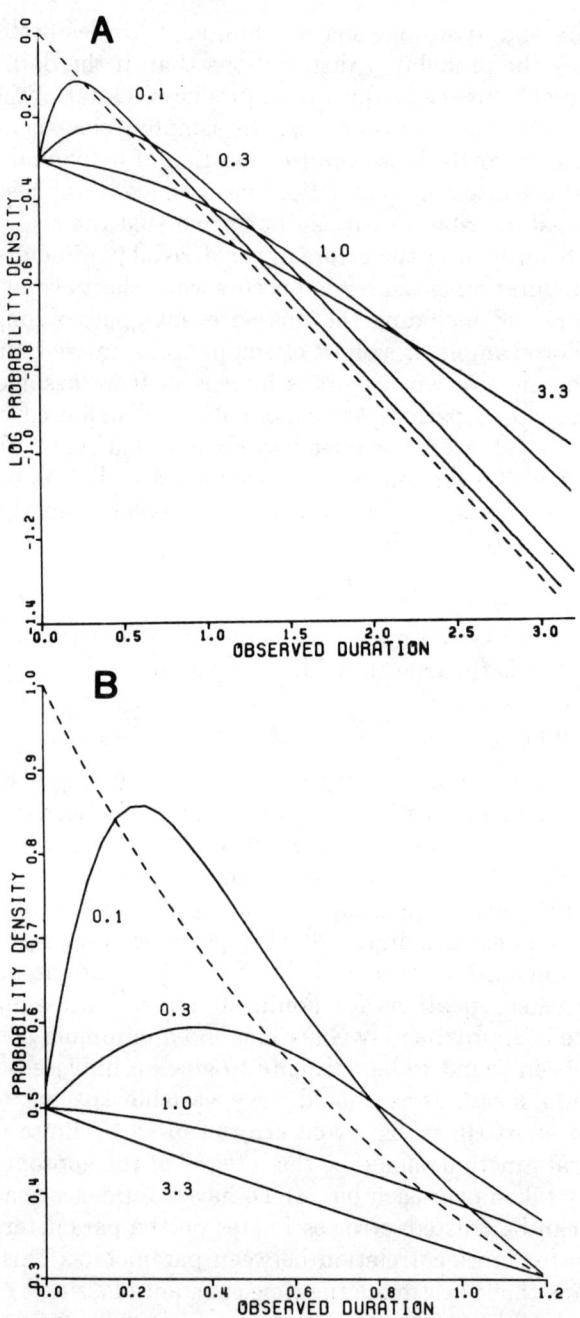

Fig. 5. *A*: distortion of exponential distribution of pulse durations produced by band limiting for noise-free system filtered by single-pole filter with time constant τ. Population distribution has time constant of 1. *Dotted line* is distribution expected for infinite

process, if the time constants are well separated (by factor of 10) and the two components appear in equal proportions, more than 5,000 events must be measured to extract time constants with <10% error with 90% confidence. Even more events are required for components that are less well separated. Such large numbers of events are usually unobtainable experimentally.

Estimating event amplitudes

If the amplitudes of the events of interest are well resolved, measuring current amplitudes presents no serious statistical problems; amplitudes may be quantified by several methods (2). In one method each data point is entered into an amplitude histogram referenced to the base line. Total-amplitude histograms can be constructed for the entire data set (Fig. 6) or for selected portions of the data such as bursts. When event durations are long relative to the system bandwidth, the total-amplitude histogram will appear as a series of roughly Gaussian peaks with means that differ by the single-channel amplitude. When multiple channels with different conductances are present, peaks will be present at the sum and difference amplitudes. Whether these peaks can be resolved or not depends on the noise level, the difference in the current amplitudes, and the duty cycle. With seriously band-limited data (i.e., leftmost channel in Fig. 1B), the total-amplitude histogram peaks will be skewed to lower values, and the location of the peak will underestimate the single-channel amplitude. Inaccuracies in the estimation of the base line will broaden the observed amplitude distribution.

A second method of determining event amplitudes is to measure transition amplitudes. Because of the short time scale involved, this method is less sensitive to inaccuracies in the base-line estimation and to the presence of independent overlapping channel currents, but it is more sensitive to the presence of subconducting states of the same channel.

Only channel currents that are steady before and after the transition can be accurately measured. An allowance for settling time must be made

---

bandwidth recording ($\tau = 0$). *Solid lines* are distributions that would be recorded for different values of τ. Number adjacent to each *curve* is ratio of t to population time constant (i.e., 0.1 indicates distribution time constant is 1/10 that of filter time constant). Time axis is in units of population time constant (i.e., duration of 1 corresponds to event with duration equal to mean duration). Note that significant deviations exist even for processes that are 10 times slower than filter time constant. Presence of noise will broaden curves and extend time that significant deviations occur. For higher order filters, using time constant of $0.3/f_c$ gives reasonable correspondance to curves shown. *B*: semilogarithmic presentation of data in *A* showing asymptotic approach of band-limited distribution to true distribution. For populations with mean durations comparable to system time constant, minimum delay time equal to three population time constants (not system time constants) is necessary before observed slope is close to unmodified exponential distribution.

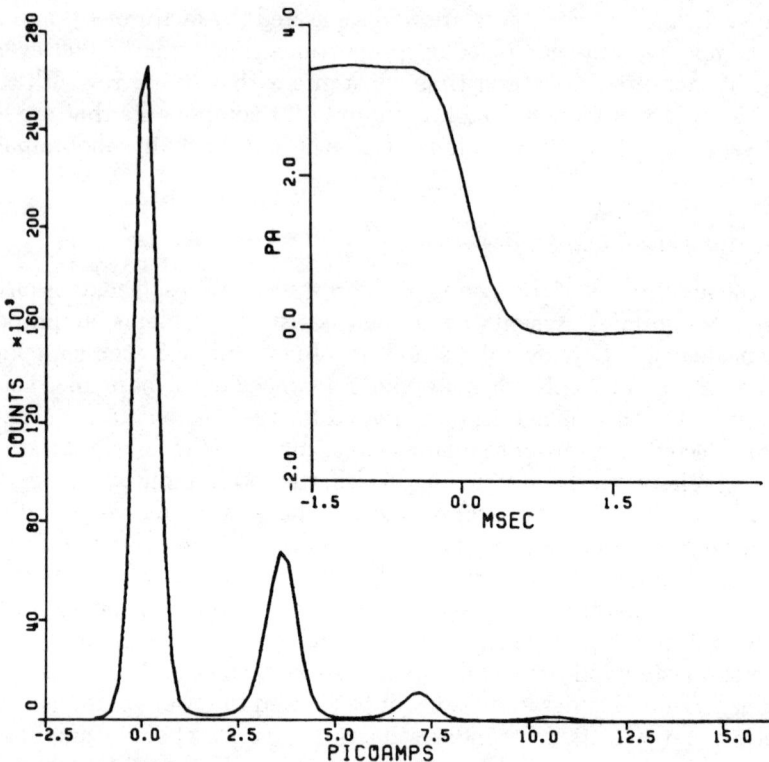

Fig. 6. Analysis of channel current amplitudes. Data are from cell-attached patch in which nicotinic channels were activated by 5 nM suberyldicholine. Base-line current level was estimated from portions of record in which no channels were open. Base-line estimate was updated approximately every 100 ms. Each data point was entered into histogram with base-line current level defined as zero. There are 4 roughly Gaussian peaks corresponding to times when 0, 1, 2, and 3 channels were open. Means are at 0, 3.50, 7.00, and 10.46 pA, respectively. Distributions are skewed to lower values because of band-limited and subconducting gaps. Relative area under each peak approximates fraction of record spent at each current level (0.70, 0.25, 0.04, and 0.01, respectively). Relative time spent at higher current levels may be underrepresented because of band-width limitations. Variance of each distribution (86, 171, 208, and 264 fA^2, respectively) increases with mean. *Inset* shows average burst trailing edge from same record (inward current is up). Bursts were defined as clusters of 1 or more open periods separated from each other by >5 ms. Burst edges were aligned at point of threshold crossing before averaging. To allow for amplifier settling, only transitions in which final open period remained above threshold for at least 250 μs were included in average. Mean transition amplitude was 3.54 pA. If continued over longer time, average current would eventually approach mean of entire record. Sampling interval was 50 μs and system bandwidth was 2.5 kHz (4-pole Bessel low-pass filter). Transition rise time is that of filter.

before and after the transition because multipole filters have a sigmoidal step response. The settling time allowance depends on the system bandwidth; two rise times is a reasonable value. In the inset of Figure 6, all burst terminations satisfying the settling-time criteria were aligned at the moment of threshold crossing and averaged. From the averaged record the mean transition amplitude is measured from one settling time before to one settling time after threshold crossing. The average current will fall on either side of the transition point because of the exponential spread in durations. However, the average transition amplitude does not give any information about the distribution of amplitudes. This can be obtained by estimating transition amplitudes on an event-by-event basis and then constructing histograms of the measured amplitudes.

A third method of characterizing amplitudes is to simply calculate the mean of the data points (after allowing for settling) within each open and/or closed period (2). Note that the number of points used to estimate the mean varies with the duration of the event, so that the amplitude of long events is estimated with greater precision than the amplitude of short events. Thus the amplitude distribution of these means is non-Gaussian. The presence of currents that do not cross threshold will bias the results. For example, open-channel currents will appear smaller because of unresolved closures.

Estimating number of active channels in patch

If the channels of interest do not desensitize or inactivate, then estimating the number of channels active in the patch is a simple matter of giving a maximal stimulus (i.e., voltage step or agonist dose), measuring the total channel current, and dividing by the single current amplitude. With channels that cannot be driven to saturation the number of channels active in the patch may be estimated by observing the fraction of time spent with 0, 1, 2, ..., n channels open (16). This discrete-level histogram is expected to follow a binomial distribution. From a nonlinear regression of the data to a linearly interpolated binomial distribution, the total number of channels that can be activated (N) and the probability of each channel being open (p) can be estimated. Alternatively, a maximum likelihood estimate of N can be made. The likelihood of n_i observations at level i is the product of the probabilities for each observation. Thus the likelihood (L) of observing a particular data set of n_i is given by

$$L = \prod_{i=0}^{i_{max}} [P(i, N, p)] n_i \tag{8}$$

where $P(i,N,p)$ is the binomial probability of i successes (current level) out of N trials (channels) and i_{max} is the maximum observed level. If L is maximized with respect to N and p, these numbers constitute the best guess for both. The maximization with respect to p can be done analytically and is simply the mean level divided by N, which has yet to be determined

$$p = \sum_{i=0}^{i_{max}} in_i / (N \sum n_i) \quad (9)$$

The maximization with respect to N using the above equation as a constraint is most simply accomplished by calculating $\ln(L)$ for $N = i_{max}$ to perhaps 50 and searching for the maximum.

Two conditions limit the accuracy of binomial calculations. *1)* When the Np product is <0.1, the distribution appears Poisson and N and p are no longer separately distinguishable. The method does not work for sparse data. *2)* The highest levels (for $p < 0.5$) will be underrepresented because of bandwidth limitations and thus will bias the estimate to lower values of N. Estimating confidence limits on N requires some knowledge of the number of experiments performed or the degrees of freedom. This number is not the number of data points or the number of levels but is approximately the number of independent opening events contributing to the data. The weighting and bin width considerations (see *Estimating event durations*, p. 142) for histograms of event durations must be followed when fitting discrete-level histograms.

Summary

The patch-clamp technique affords unparalled resolution of the detailed properties of ion-channel currents. Patch clamping is not difficult and has been used to record single-channel currents from many cell types. The number of artifacts associated with the method appears to be rather small. The main experimental difficulties arise from the need to process large amounts of data. With the development of inexpensive computers and mass storage devices, these problems should be alleviated. The study of membrane excitability is expanding rapidly due to the introduction of high resolution patch-clamp techniques. The conceptual revolution that will inevitably follow is just beginning.

We thank J. Neil, R. McGarrigle, F. Guharay, J. Hidalgo, R. Spangler, and T. Smith for their advice and comments on the manuscript. This work was supported by NS-13194 from the USPHS NINCDS.

Major portions of this work have been reproduced from: Sachs, F., and A. Auerbach. Single channel electrophysiology: use of the patch clamp. In: *Methods in Enzymology: Neuroendocrine Peptides*, edited by M. Conn. New York: Academic, 1983, p. 147–176.

REFERENCES

1. Aldrich, R. W., and G. Yellen. Analysis of nonstationary channel kinetics. In: *Single-Channel Recording*, edited by B. Sakmann and E. Neher. New York: Plenum, 1983, p. 287–299.

2. Auerbach, A., and F. Sachs. Single channel currents from acetylcholine receptors in embryonic chick muscle: kinetic and conductance properties of gaps within bursts. *Biophys. J.* 45: 187–198, 1984.
3. Ayer, R. K., Jr., R. G. DeHaan, and R. Fischmeister. Measurement of membrane patch and seal resistance with two patch electrodes (Abstract). *J. Physiol. London* 34: 37P, 1983.
4. Bevington, R. P. *Data Reduction and Error Analysis for the Physical Sciences.* New York: McGraw-Hill, 1969.
5. Colquhoun, D., and A. G. Hawkes. On the stochastic properties of single ion channels. *Proc. R. Soc. London Ser. B* 211: 205–235, 1981.
6. Colquhoun, D., and A. G. Hawkes. On the stochastic properties of bursts of single ion channel openings and clusters of bursts. *Proc. R. Soc. London Ser. B* 300: 1–59, 1982.
7. Colquhoun, D., and F. J. Sigworth. Analysis of single channel data. In: *Single-Channel Recording*, edited by B. Sakmann and E. Neher, New York: Plenum, 1983, p. 191–263.
8. Dionne, V. E., and M. D. Leibowitz. Acetylcholine receptor kinetics: a description from single channel currents at snake neuromuscular junctions. *Biophys J.* 39: 253–261, 1982.
9. Fenwick, E., A. Marty, and E. Neher. A patch-clamp study of bovine chromaffin cells and of their sensitivity to acetylcholine. *J. Physiol. London* 331: 557–597, 1982.
10. Fenwick, E., A. Marty, and E. Neher. Sodium and calcium channels in bovine chromaffin cells. *J. Physiol. London* 331: 599–635, 1982.
11. Guharay, F., and F. Sachs. Stretch-activated single ion-channel currents in tissue-cultured embryonic chick skeletal muscle. *J. Physiol. London.* In press.
12. Hamill, O. P., A. Marty, E. Neher, B. Sakmann, and F. J. Sigworth. Improved patch-clamp techniques for high-resolution current recording from cell and cell-free membrane patches. *Pfluegers Arch.* 391: 85–100, 1981.
13. Horn, R., and K. Lange. Estimating kinetic constants from single channel data. *Biophys. J.* 43: 207–223, 1983.
14. Lux, H. D., E. Neher, and A. Marty. Single channel activity associated with the calcium dependent outward current in Helix pomatia. *Pfluegers Arch.* 389: 293–295, 1981.
15. Numann, R., R. K. S. Wong, and R. B. Clark. Electrophysiology of single dissociated cortical neurons. *Soc. Neurosci. Abstr.* 8: 413, 1982.
16. Patlack, J., and R. Horn. Effect of N-bromoacetamide on single sodium channel currents in excised membrane patches. *J. Gen. Physiol.* 79: 333–351, 1982.
17. Sachs, F., J. Neil, and N. Barkakati. The automated analysis of data from single ionic channels. *Pfluegers Arch.* 395: 331–340, 1982.
18. Sachs, F., and P. Specht. Fast microelectrode headstage for voltage clamp. *Med. Biol. Eng. Comput.* 19: 316–320, 1981.
19. Sakmann, B., and E. Neher (editors). *Single-Channel Recording.* New York: Plenum, 1983.
20. Sieglebaum, S. A., J. S. Camardo, and E. R. Kandel. Serotonin and cyclic AMP close single K^+ channels in *Aplysia* sensory neurones. *Nature London* 299: 413–417, 1982.
21. Trautmann, A. Curare can open and block ionic channels associated with cholinergic receptors. *Nature London* 298: 272–275, 1982.
22. Trautmann, A., and S. A. Siegelbaum. The influence of membrane isolation on single acetylcholine-channel current in rat myotubes. In: *Single-Channel Recording*, edited by B. Sakmann and E. Neher. New York: Plenum, 1983, p. 473–480.

SEVEN

Voltage Clamp and Internal Perfusion With Suction-Pipette Method

Arthur M. Brown
David L. Wilson
Yasuo Tsuda

*Department of Physiology and Biophysics,
University of Texas Medical Branch, Galveston, Texas*

Design Considerations • Practical Tests of Adequacy of Exchange • Voltage Clamp • Time Resolution and Spatial Control of Voltage Clamp • Circuit and Cell Characteristics • Preparation of Medium-Sized Neurons • Discussion • Summary

---

The method discussed in this chapter fits between perfusion methods applied to very large cells such as the squid axon (1, 21) and diffusion methods applied to very small cells such as chromaffin cells (9). The suction-pipette method has been applied to spherical cells such as snail neurons between 50 and 100 μm in diameter (19) and to cylindrical cells such as mammalian ventricular myocytes with diameters $\geqslant 10$ μm and lengths $\leqslant 200$ μm (1). There are several similar techniques available for the study of cells with these dimensions. Kostyuk et al. (15) originally introduced a partition method to isolate single neurons and have subsequently extended the method to a form of suction pipette (16). Takahashi and Yoshii (22) and Byerly and Hagiwara (5) also used modified suction-pipette methods to achieve the same results. In squid axon true perfusion is performed; in chromaffin cells exchange with intracellular components is by diffusion from a micropipette. In intermediate-sized cells, such as snail neurons in which the suction pipette is useful, control of the milieu interieur is achieved mainly by dialysis, although some bulk flow may occur. A true flow-through system based on simultaneous use of two suction pipettes has also been described (2, 18). In this chapter the adequacy of the exchange and the limitations of the voltage-clamp system are emphasized, beginning with an account of the design and fabrication.

Design Considerations

There are two important features of design that may conflict with each other. *1)* The shunt resistance (R_{sh}) must be larger than the input resistance of the cell (R_{in}) and *2)* the internal perfusion must be satisfactory. However, a good seal (high R_{sh}) is most often established when a small suction pipette is used, whereas fast exchange occurs with a large pipette. For cells 50–100 µm in diameter, acceptable $R_{sh}:R_{in}$ ratios of 10 or greater can be obtained with inner tip diameters of 20–30 µm and substantial exchange of K^+ and Cl^- can occur within 1 min of perfusion (Fig. 1). Tris was substituted for K^+ and aspartate was substituted for Cl^- in the internal perfusion fluid; the rate of change of K^+ or Cl^- was measured by the appropriate liquid ion–exchanger microelectrode. Both ion-selective electrodes have high selectivities against the substituted ions. As Figure 1 indicates, exchange was quick and nearly complete. Exchange of K^+, Cl^-, and Na^+ is reasonable, but control of Ca^{2+} levels is not complete (see **Practical Tests of Adequacy of Exchange**, p. 155). With smaller cells (heart cells or small neurons) pipettes >10 µm in diameter may be required to achieve acceptable $R_{sh}:R_{in}$ ratios. In such cases perfusion is not as quick, although the smaller cell volume reduces the requirement for speed of perfusion. When perfusion needs to be enhanced, the pipette should taper as steeply as possible (cf. Fig. *2a-1* and *2a-2*). This allows one to place the inlet tube very near the pipette tip and thus enhance exchange. As noted, the term perfusion may not be accurate because significant bulk flow through the cell may not actually occur. However, in our approach the membrane is ruptured by a wire and bulk flow under a pressure head does occur near the rupture. The process is probably a mixture of bulk flow and dialysis, but for the sake of simplicity the term *perfusion* is used.

The pipettes are pulled from Pyrex tubing (outer diam, 3.0 mm) and tapered over a length of 1.0–2.5 cm from the tip (Fig. 2). The shank is scored with a diamond knife depending on the internal tip diameter required. For a diameter of 30–50 µm, the score can be made ~1.0 cm from the tip. The electrode is then broken, and the break must be perfect—no jagged edges. Fire polishing and addition of a hydrophobic coating of a 50:50 mixture of mineral oil and Vasoline are then done as shown in the figure. The hydrophobic coating is not essential, however.

The pipette assembly is shown in Figure 3. This particular arrangement is not crucial, but several features are important. *1)* The inlet tubing must be extended as close to the tip of the suction pipette as possible after the pipette has been added to the assembly (Fig. 3); this is important for speed of exchange. *2)* The carrier wire to which the puncture wire is attached must be easily moved. *3)* The side chamber containing the potential electrode must be baffled (we use gauze and wooden stem of

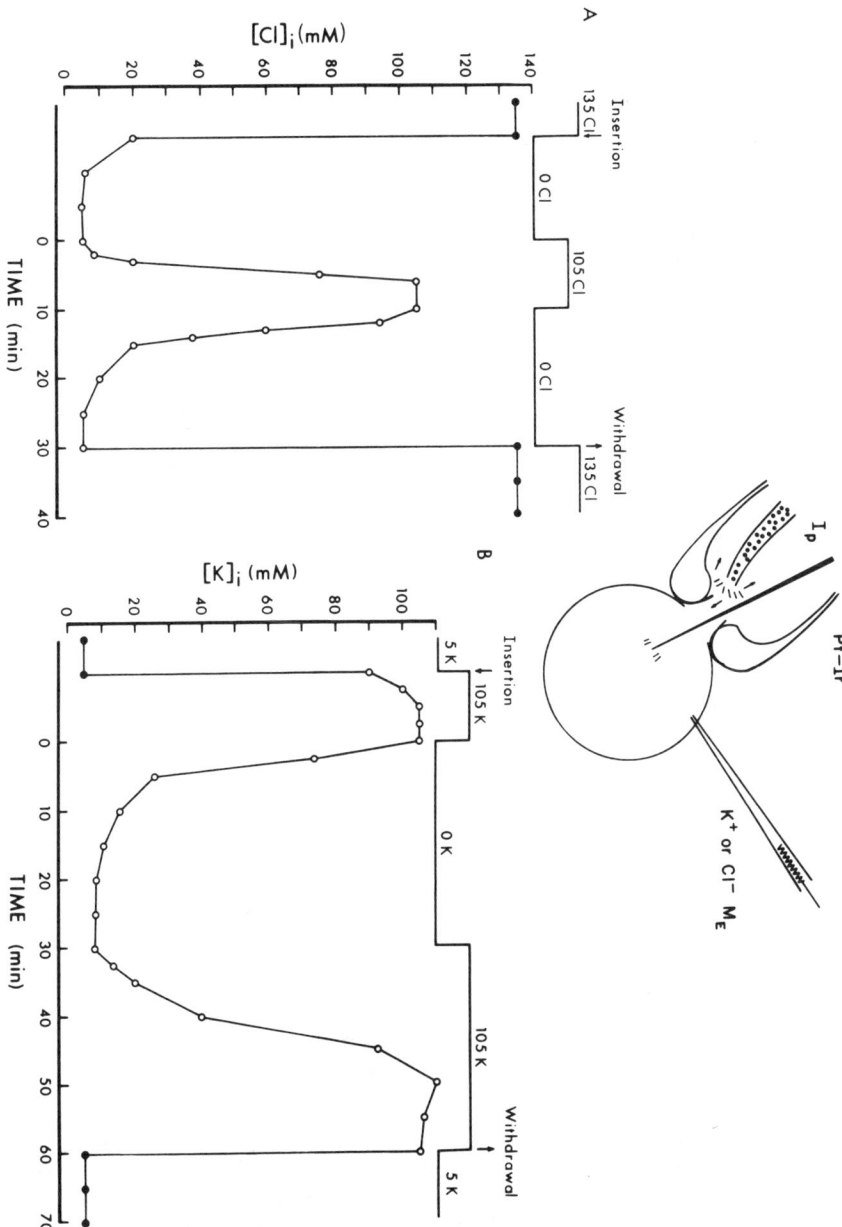

Fig. 1. *A*: changes in intracellular Cl⁻ concentration after changes in internal perfusate Cl⁻. Concentration calculated from measured ionic activities with measured activity coefficient of 0.77. The Cl⁻ microelectrode is inserted (*down arrow*) into neuron perfused with potassium aspartate. Then 105 mM Cl⁻ was substituted for aspartate ion. ○, extracellular Cl⁻ activities; ●, intracellular Cl⁻ activities. [From Lee et al. (14).] *B*: changes in intracellular K⁺ concentration after changes in internal perfusate. $Tris_i^+$ was substituted for K_i^+; sequences as in *A*.

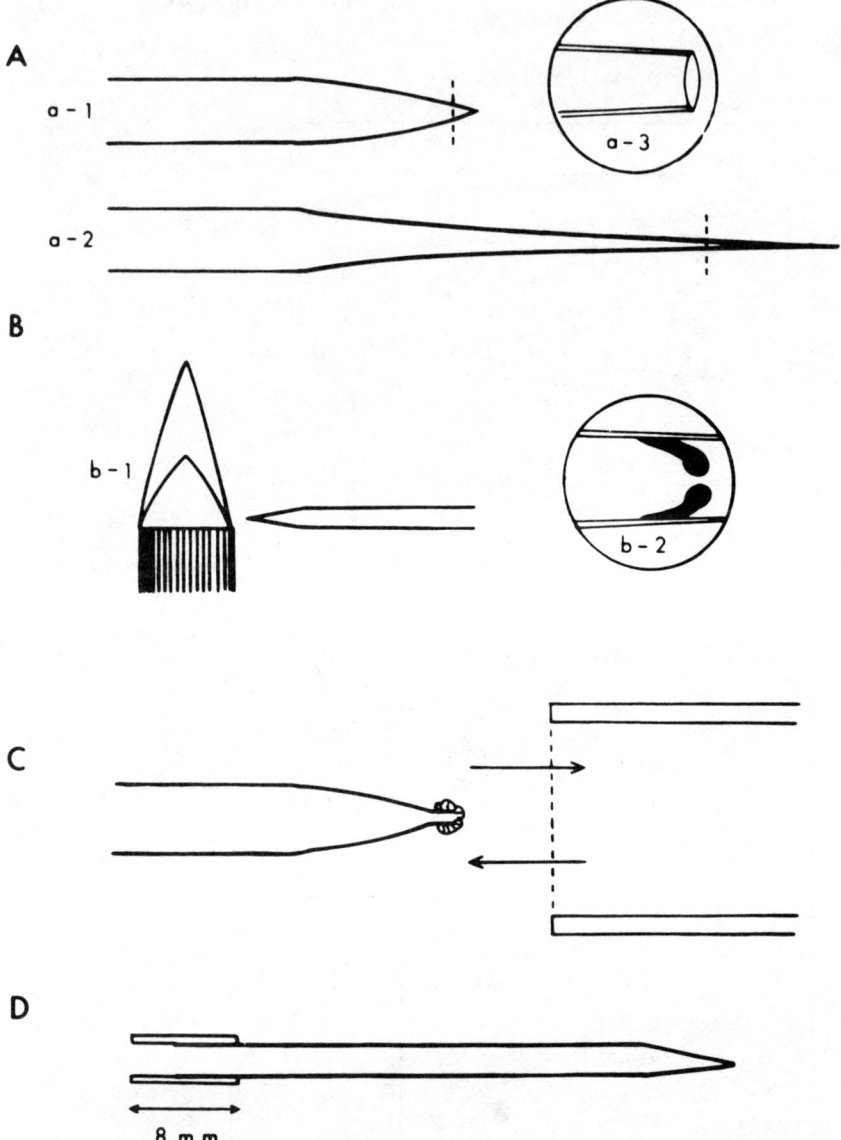

Fig. 2. Manufacture of suction pipette. *A*: 2 types of suction pipette. *a-1*, Short-shank type used for rapid perfusion; *a-2*, long-shank type in which perfusion is slowed; *a-3*, orifice after fracture of shank at *dotted lines* in *a-1* and *a-2*. Electrodes are matched to needs of experiment: short-shank electrode is used for applying several solutions successively. Better seals are obtained with short-shank type because surface area of cell in contact with glass is decreased. *B*: fire polishing of fractured tip. *b-1*, Fractured tip is fire polished in gas flame; *b-2*, enlarged tip orifice before and after (*heavy lines*) fire-polishing. *C*: Vaseline coating of tip. After dipping tip into small amounts of Vaseline, it is inserted into large heated glass tube resulting in hydrophobic coating. *D*: fitting of vinyl tube to act as pipette gasket connecting pipette to antechamber.

cotton applicator stick) to prevent pressure fluctuation from affecting the electrical measurements. Full details of the method of fabrication have been presented by Lee et al. (20).

Liquid junction potentials (LJPs) between the usual pipette solutions and the extracellular solution were <2 mV (19). For the more exotic solutions they must be measured each time. Because we generally use a separate micropipette filled with 3 M KCl as a voltage probe, this becomes the reference for the measurement of LJPs. When we change the internal perfusate from K-aspartate to Cs-aspartate, the measured LJP change is ~2 mV.

Practical Tests of Adequacy of Exchange

It is not convenient to insert ion-selective microelectrodes into each cell to establish the extent of the exchange. When the suction pipette is used on snail neurons, the adequacy of internal perfusion for the isolation of Ca currents is routinely assessed with the following criteria. With a normal-Ringer external solution present, internal perfusion of the aspirated cell is begun with an internal solution in which K^+ has been replaced by Cs^+. In ~10–20 min the duration of action potentials produced by short current pulses gradually increases from a half-peak value of ~20 ms to 3–4 s. After this occurs the external bath is changed from normal-Ringer solution to a solution in which Na^+ has been removed and K-current (I_K) blockers have been added (3); this causes no significant change in the action-potential duration. By contrast, when K^+ rather than Cs^+ is present in the internal solution, changing the external solution has large effects on the action-potential duration.

Internal perfusion with EGTA [ethylene glycol-bis(β-aminoethylether) N,N-tetraacetic acid] is another way to test intracellular exchange. The Ca^{2+}-current–dependent component of Ca-current (I_{Ca}) inactivation is attenuated by this procedure. Figure 4A shows that in the presence of 3 mM $EGTA_i$ the peak of I_{Ca} is slightly increased and the subsequent reduction of I_{Ca} is slowed. The inactivation of I_{Ca} can be fit by the sum of two exponentials (3), and Figure 4B shows that the faster time constant of inactivation is primarily slowed by $EGTA_i$.

However, Ca buffering is not complete. Evidence for this comes from observations such as those shown in Figure 5, where $EGTA_i$ had been added previously to the internal perfusate and the rate of Ca-current inactivation slowed (Fig. 4). The buffering is not complete, however, because the rate of inactivation slowed further when Ba^{2+} (which does not produce inactivation) was substituted for Ca^{2+} extracellularly. More direct measurements of intercellular Ca with a liquid ion–exchanger Ca electrode under conditions in which buffered Ca solutions were exchanged have also demonstrated the difficulty of Ca buffering (6).

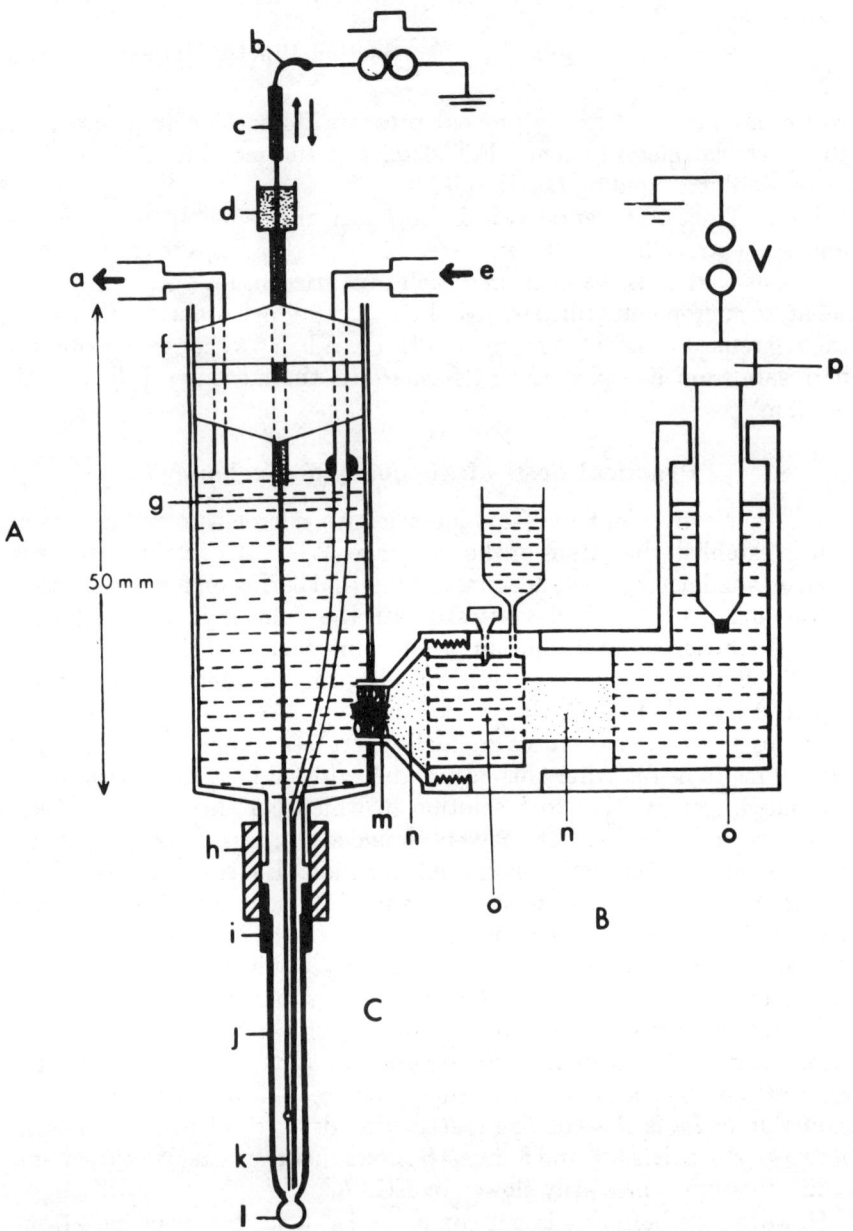

Fig. 3. Suction pipette. A: antechamber. B: attachment for recording membrane potential. C: suction-pipette attachment. a, Outlet tube for suction and internal perfusate; b, Ag wire to apply constant current during current clamp and feedback current during voltage clamp; c, portion insulated by polyethylene tube allowing puncture wire to be hand driven; d, guide filled with silicone grease or Vaseline for Ag wire; e, inlet tube for internal perfusate; f, rubber plug; g, tapered polyethylene tube for inflow; h, polyethylene gasket connecting suction pipette to antechamber; i, vinyl fitting; j, glass suction pipette; k, puncture wire (90% Pt platinized, 10% Ir) soldered to Ag wire and sharpened electrolytically to 1–2 μm tip diam; l, single cell; m, connection between calomel half-cell and antechamber; n, Ringer–1%-Agar bridge; o, 0.15 M KCl; p, calomel half-cell. [From Lee et al. (14).]

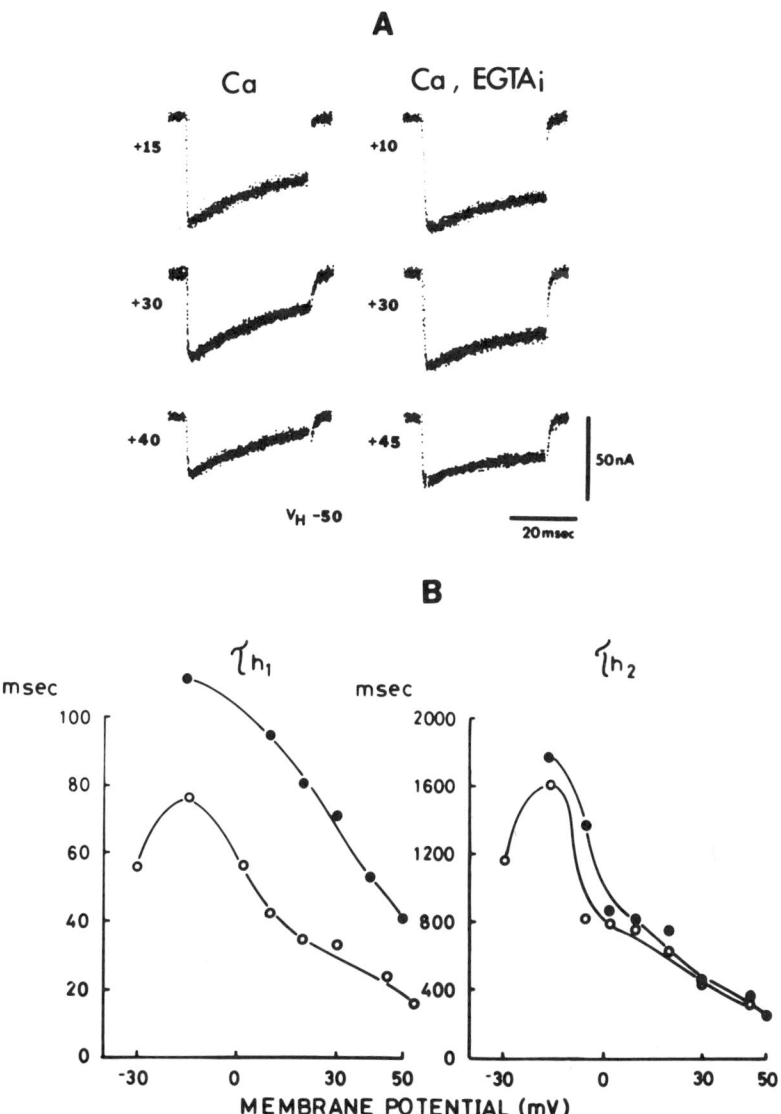

Fig. 4. Effect of internal perfusion with EGTA on inactivation of Ca current. Note that in presence of EGTA, inactivation present in records decreases. Longer voltage than those shown in A displayed a biexponential decay of current. These were fit with sum of two exponentials, and resulting time constants (τ_{h_1} and τ_{h_2}) are shown in B. Measurements were obtained on cell before (○) and after (●) internal perfusion with EGTA.

Voltage Clamp

In large cells in which the electrode resistance approaches the input resistance of the cells, it is often impossible to achieve satisfactory

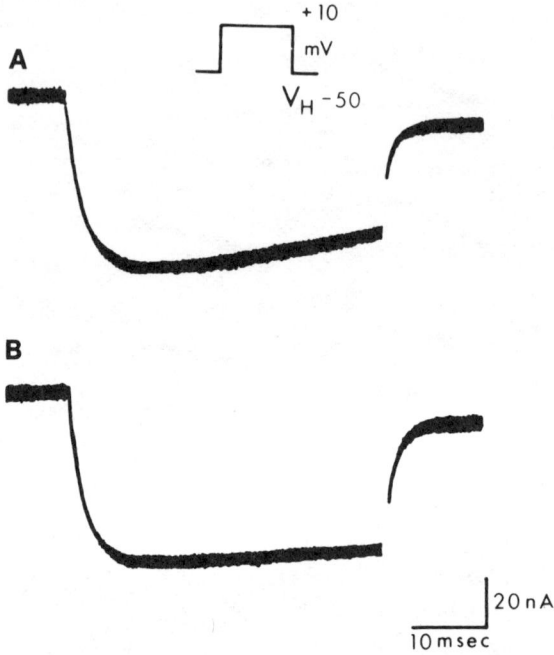

Fig. 5. Evidence that perfusion with $EGTA_i$ does not completely buffer Ca^{2+} influx. Prior to record, 3 mM $EGTA_i$ was perfused intracellularly for 50 min producing effects similar to those seen in Fig. 4. Substitution of Ba^{2+} for Ca^{2+} extracellularly further reduced rate of inactivation.

potential control with a single-electrode clamp because of the access resistance of the pipette. This access resistance is effectively a large series resistance, which is almost impossible to compensate completely (10). Thus a hybrid clamp consisting of a suction pipette for passing current and a microelectrode for sensing the voltage is used. This is referred to as a combined voltage clamp (Fig. 6). The clamp amplifier is a 1030 Teledyne Philbrick with a gain-bandwidth product of 100 MHz and a high-frequency rolloff that allows high stability. Because of the latter feature, circuit stability can be maintained at a relatively high frequency, even in the presence of the resistance-capacitance (RC) time constant of the cell (10). The glass microelectrodes for measuring membrane potential were painted with silver, insulated, and driven with capacitance compensation. This resulted in a significant increase in the bandwidth of the voltage measurement, and in several cases we found

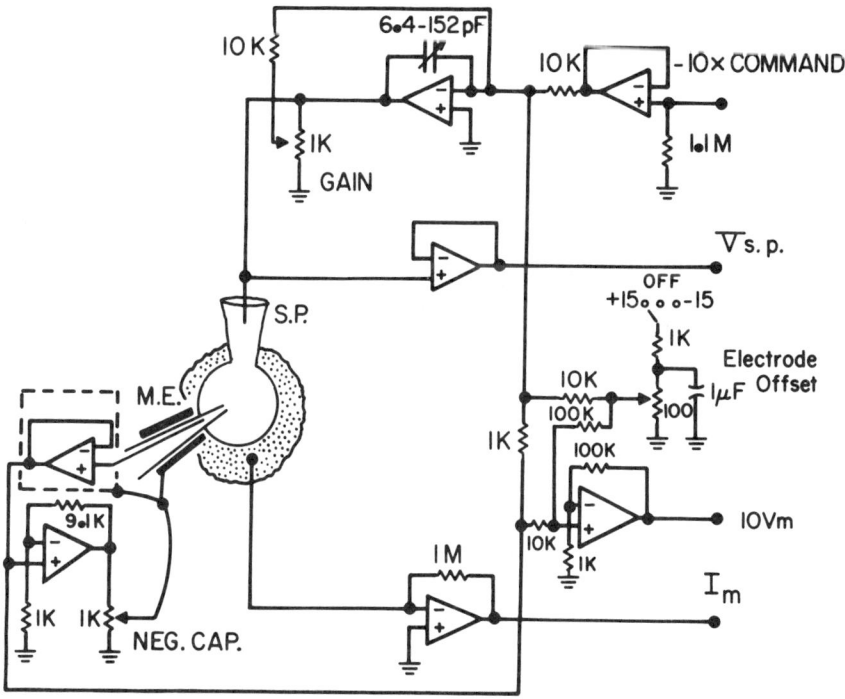

Fig. 6. Voltage-clamp system. Microelectrode (ME) was used for voltage sensing and suction pipette (SP) was used for passing current and internal perfusion of cell; they are shown with neuron immersed in patch of fluid. Microelectrode headstage was attached directly to microelectrode holder; headstage and microelectrode shield were driven from capacitance-neutralization circuit. In addition to circuit components shown, design included other features such as current pump source, model cell circuit, and circuit for testing microelectrode response. All op amps were LF356s except voltage-clamp amplifier (with variable capacitor), which was Teledyne-Philbrick 1030. (Full circuit details are available on request.)

that the amplitude response of carefully painted and compensated microelectrodes was flat up to as much as 10 kHz. The inlet and outlet tubing surrounding the suction electrode were shortened to reduce stray capacitances. Nevertheless a measurement of the current in the bath was necessary because there was significant capacitive shunting of the suction-pipette apparatus to ground; thus current measurements were made with a current-to-voltage (I-V) converter circuit (Fig. 6); the bath electrode was an Ag/AgCl pellet. After application of voltage clamp, the circuit was tuned so that its full gain was applied to the circuit, and the compensation capacitor was set at the minimum setting to obtain a damped voltage trace.

Time Resolution and Spatial Control of Voltage Clamp

The capacitive transient decays in two phases (Fig. 7A), and the large fast phase subsides almost totally within 60 μs. A slower phase of the relaxation (amplitude, ~5% of fast phase) was also present as reported by Byerly and Hagiwara (5). The nature of this slow phase is unknown, although Byerly and Hagiwara (5) suggested a relationship to the piece of membrane aspirated in the suction pipette. Preliminary measurements indicate a constant phase angle in impedance measurements similar to that observed in squid axon, and this is consistent with such a capacitive transient (7). Over the voltage range examined, linearity of both components was confirmed by scaling and overlaying hyperpolarizing responses. These linear capacitive components of the response were removed by summation of equal but opposite voltage-clamp pulses.

The spatial control was examined with a second, externally silver-painted, insulated microelectrode with negative capacitance compensation that was "outside" the clamp circuit. The suction pipette and the two glass microelectrodes were placed at three points forming a roughly triangular distribution about the cell. Conditions were imposed that placed the greatest stress on spatial control, i.e., control during tail currents produced by steps to potentials that activate I_{Ca} maximally. Figure 7B and C shows the voltage differences between the two microelectrodes ($V_1 - V_2$) at these times. The difference is ~3 mV within 30 μs and ≤1 mV within 70 μs; this measurement includes errors such as the imperfect matching of microelectrodes. We conclude that the spatial voltage control of the cell interior is complete within 50 μs under conditions of maximum stress. Summed hyperpolarizing and depolarizing voltage traces from the voltage-clamp microelectrode were monitored; the maximum deviation from a straight line was ~0.03 mV. This indicates that the voltage pulses were symmetrical and that there was no loss of control because of ion conduction. The linear components of the capacitive current transient are thus readily removed by subtraction, allowing the fast tail currents to be separated.

These results suggest that the internal resistance of the cell is small and that the almost spherical neuron is spatially clamped quite well in the bulk of the cell interior. Nevertheless the presence of a series resistance (R_s) near the cell membrane would still impede the analysis of the membrane properties by creating an ohmic voltage drop between the cell interior and the extracellular electrode. In the next section the series resistance of the snail preparation and some of the other characteristics of the voltage-clamp system used in our lab are evaluated.

Circuit and Cell Characteristics

Preliminary measurements of the impedance of the snail neuron were made to evaluate R_s. The measurements were conducted with a system

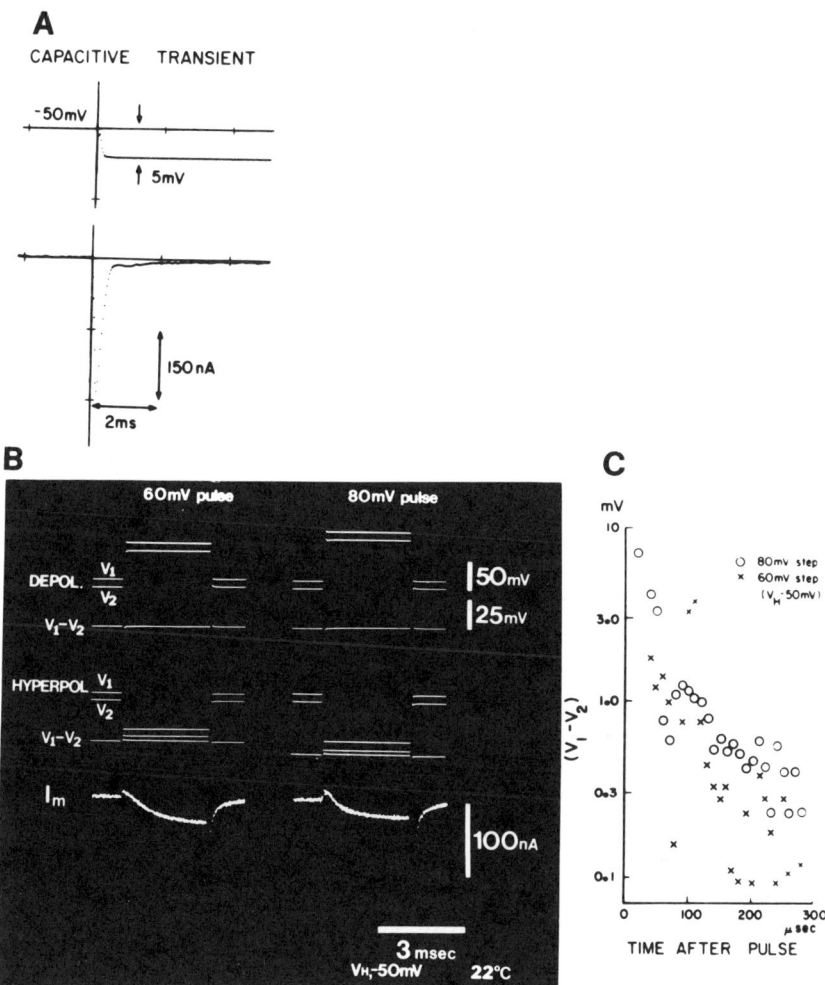

Fig. 7. Capacitive transient and isopotentiality of cell during voltage clamp. *A*: capacitive transient for subthreshold depolarizing pulse. Voltage trace has rise time of ~25 μs. There were 2 phases in decay of current: faster phase was almost complete in 60 μs, slower much smaller phase lasted >0.5 ms. *B*: isopotentiality of cell body during voltage clamp. Voltages V_1 and V_2 are measured by different microelectrodes, one of which was used in voltage-clamp circuit (see Fig. 6); V_2 is displaced vertically with respect to V_1. Voltage steps were depolarizing steps of 60 and 80 mV; difference between voltages with smaller steps was much less. Voltage difference between two electrodes is also displayed. Currents result from summation of equal and opposite voltage pulses. Note that current traces show initial outward-asymmetry current and that tail currents relax in two phases. Some initial data points in tail currents have not reproduced in this figure; size of initial tail currents are ~200–300 nA. *C*: voltage difference ($V_1 - V_2$) is displayed on logarithmic scale. Voltage difference was <1 mV at 100 μs.

developed by Fishman (10) for making rapid transfer-function measurements. The system consists of a pseudorandom noise source (PS) and a spectrum analyzer (Rockland 512/s) connected to an LSI-11/23 microcomputer (11, 12). There are several different Fourier-synthesized PS waveforms stored in a read-only memory; this allows one to select a waveform that applies a spectral content that will give good signal-to-noise characteristics at all frequencies in the output signal. Fourier transforms of the input and output were obtained on 1,024 digitized points, and 400 of the frequency domain points were displayed. A low-amplitude PS signal was applied to a current pump circuit (10) for passing current into the cell interior via the suction pipette. A low-impedance microelectrode (1 MΩ) with a driven shield for capacitance compensation was used as the voltage-sensing element. Current was collected from an Ag/AgCl pellet in the bath, and an I-V circuit was used to measure the current.

The input signal for the measurement of the cell impedance was the current signal derived from the I-V converter. Note that this signal was a true measure of the current through the cell, whereas the PS signal was not correct because of capacitive shunting of the current from the current pump circuit to ground via paths other than through the bath, as described below. The output signal for the measurement of the transfer function was the voltage in the cell interior. Two corrections were necessary to make the measurements because 1) the frequency response of the microelectrode system was limited and 2) the I-V converter presented an input impedance to ground that looked like an ideal inductor and that became significant at ~5–6 kHz. For each measurement, the current-to-voltage transfer function was obtained and at each frequency value a complex division of this value by a value obtained from the measured microelectrode transfer function was performed. The complex I-V converter impedance was subtracted from this result to obtain the cell impedance.

Certain other components of our measurement system have been evaluated. The frequency response of the microelectrode system was measured with the same arrangement shown in Figure 8A except that the cell was removed and a 500-kΩ resistor was placed between the pick-up electrode and the I-V converter. Thus the current signal can be simply scaled by the 500-kΩ resistor to obtain the voltage in the bath. Using 1-MΩ microelectrodes with carefully adjusted capacitance compensation, we obtained a magnitude response flat to near 10 kHz (Fig. 8B). We found that there was significant capacitive shunting of the suction pipette to ground via paths other than through the bath, as measured by the following method. With the suction pipette in the bath, a current measurement, obtained from the voltage drop across a resistor that connects the current pump circuit to the suction pipette, was compared with the current measurement from the *I-V* converter. There was a difference in these two measurements at high frequencies that appeared to be caused by ~40 pF of capacitance shunting the suction-pipette apparatus to ground. Thus all measurements were made with the bath electrode and the *I-V* converter. In addition to this measurement we wanted to evaluate the capacitive shunting through the suction-pipette glass into the bath. This current component would circumvent the cell membrane but be recorded by the bath electrode; thus it presented a possible source of error. This capacitive shunting was estimated by measuring the current in the bath when the suction pipette was driven from the current pump and the tip of the suction pipette was occluded. This measurement indicated a small capacitive component, roughly estimated to be <3 pF that should therefore have negligible effects on the measurement.

The results for a cell with the normal intra- and extracellular solutions are shown in Figure 8C (3). The high-frequency asymptote of the impedance is expected to yield the series resistance of the cell. In our case the impedance was not purely resistive at the highest frequency displayed; R_s is thus estimated to be <5 kΩ. Note that the phase lies between −60° and −85° over a wide frequency range and that the slope of the magnitude curve is slightly <20 dB/decade; this is characteristic of a constant phase-angle capacitance as measured in squid axon (7, 10). The slow relaxation noted by Byerly and Hagiwara (5) can be explained by the constant phase angle and may be a general membrane property rather than the result of the membrane being attached to the inner wall of the pipette as suggested. A comprehensive model of the cell impedance awaits further measurements and analyses, but for our purposes R_s is sufficiently small so that it produces negligible error in the voltage clamp (~0.5 mV for peak current of 100 nA and ~1 mV during measured tail current).

Generally the measurement of tail currents after voltage steps that fully activate the current should stress a voltage-clamp system maximally because the driving force can be made large and the currents are therefore very substantial. In Figure 9 we show Ca tail currents measured with the combined voltage-clamp system. *Panel A* shows uncorrected current traces and the voltage is measured from the voltage-clamp microelectrode. *Panel B* shows currents obtained after summation of hyperpolarizing pulses of equal magnitude to correct for linear capacitive currents. Note the biexponential behavior of the relaxation. A quantitative description of the Ca tail currents has been given by Brown et al. (4).

Preparation of Medium-Sized Neurons

The circumesophageal ganglia were dissected from *Helix aspersa*, and the surrounding connective tissue was removed with fine tweezers and scissors. The ganglia were continuously perfused in normal extracellular

A
Transfer
Function
Analyzer
PS
SP
μel
Preamp
IMΩ
Input
Output
B
μel FREQUENCY RESPONSE
FREQUENCY (hz)
100
1K
10K
100K
MAGNITUDE (dB)
0
-10
-20
PHASE (deg)
-45
-90
-120
C
FREQUENCY (hz)
1
10
100
1K
10K
-1M
100K
10K
3
|Z| (Ω)
PHASE (deg)
0
-30
-60
-90

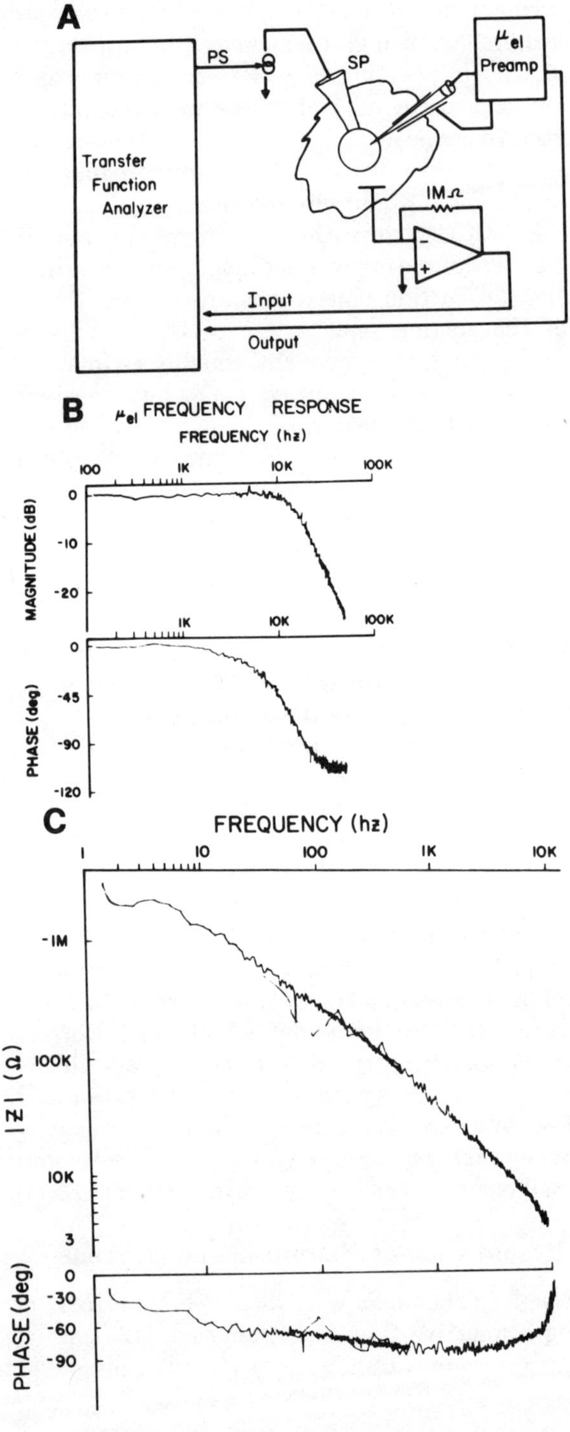

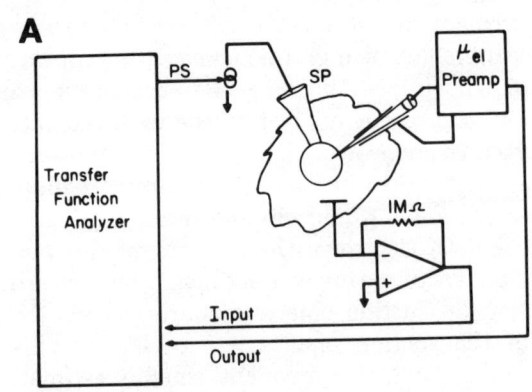

snail Ringer solution for ~15–20 min, which helps remove connective tissue. The neurons were then isolated by dissection alone. Neurons of 50–100 μm in diameter were selected because this size was large enough to accommodate our two-electrode voltage clamp yet small enough to help ensure the spatial control of the voltage. The neuron was held to the suction pipette by gentle suction (20) and the portion of membrane aspirated into the pipette was ruptured with an etched Pt-Ir wire; this step was essential for satisfactory internal perfusion. The axon was then severed near the neuron (within 50 μm) by pinching it with fine forceps or by cutting it with the tip of a broken microelectrode. When the axon was cut, the voltage response to hyperpolarizing current pulses decreased by ~20%. When the membrane impedance did not return to its original value in ~5–10 min, indicating a "healing over" of the neuron, the cell was discarded.

Three additional criteria were required before an isolated cell was used. *1)* The action-potential amplitude had to be at least 80 mV when measured with the suction pipette, and it had to remain constant during repetitive firing. *2)* The resting potential had to be about −50 mV. *3)* After impalement with a separate microelectrode the holding current required to keep the cell at −50 mV had to be $<10^{-9}$ A. Cells that met these requirements had input resistances of ~5–10 MΩ; these values were comparable to values measured after the insertion of two microelectrodes (3). Successful isolations were obtained with these methods ~25%–30% of the time.

We also tried to isolate cell bodies by exposure to various amounts of trypsin (wt/vol, 1%, 0.5%, 0.2%) in normal Ringer for ~5–10 min at room temperature. The neuron was almost completely exposed before trypsin was applied. Cells treated in this way had a high input resistance, which is characteristic of a good seal between the pipette and cell membrane. However, we found such cells to be fragile; the currents often ran down within 15–30 min after the commencement of internal perfusion. The Ca currents became small or insignificant, and leakage resis-

Fig. 8. Impedance measurements. *A*: set-up for measuring impedance. Current was applied from current pump circuit to cell via suction pipette (SP) and voltage was measured with shielded and compensated microelectrode (μ_{el}). Current was collected from bath; *I-V* converter circuit and voltage signal from μ_{el} formed input and output signals from which transfer-function analyzer obtained impedance measurement. (See text for further details.) *B*: frequency response of 1-MΩ microelectrode with capacitance compensation using carefully painted microelectrode. (See text for details of measurement.) *C*: magnitude and phase of impedance of snail neuron. Corrections applied as described in text. Measurements taken from two frequency ranges with region of overlap. High-frequency asymptote of impedance should reflect series resistance of the cell. Measured phase and magnitude do not indicate pure resistance; this suggests that series resistance is less than the minimum impedance value and is probably significantly <5 kΩ. Phase lies between −60° and −80° over wide range of frequencies, and slope of magnitude curve is slightly <20 dB/decade. This behavior indicates a constant phase-angle capacitance.

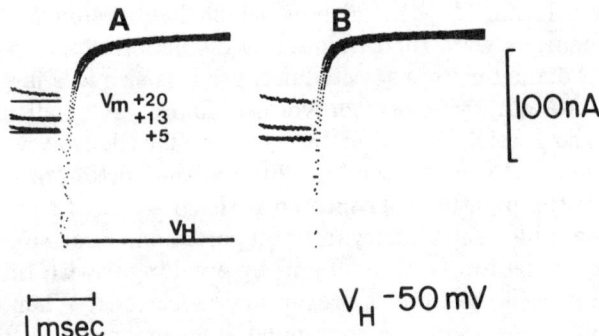

Fig. 9. Calcium tail currents. *A*: currents obtained on return to holding potential (V_H) of −50 mV after depolarizing pulses of membrane potential (V_m) shown. *Straight lines*, membrane potential recordings. Some points in tail currents fell below portion displayed here. *B*: same currents as in *A* after correction for linear capacitive currents by summing currents with equal hyperpolarizing pulses. Sums of membrane potential were just straight lines (not shown).

tance fell to very low levels. This is not to say that satisfactory enzyme treatment cannot be achieved. The experiments from the Kostyuk laboratory and the Hagiwara laboratory are evidence of this.

Discussion

The suction-pipette method, as originally envisioned, was limited to cells 50–100 µm in diameter and was used exclusively for intracellular clamping. In successful experiments, seals with resistances as great as 100 MΩ could be achieved. The gigaseal method, in which suction is applied to 1-µm-diam micropipettes pressed against cells with specially treated surfaces (13), proved that much greater seal resistances could be attained. For smaller cells, measurement of macroscopic currents with the "whole-cell" patch-clamp technique may be preferred to the suction-pipette method because better seals are obtained with a patch-type electrode (see the chapters by Auerbach and Sachs and by Lecar and Smith). However, a major requirement is that both the cell and the currents be small, otherwise capacitive or active current flow will result in a significant voltage drop across the access resistance of the pipette. Generally this access resistance will result in voltage errors if uncorrected or in a decreased bandwidth as required to stabilize a circuit with an analog compensation scheme.

For medium-sized cells the suction-pipette method offers some advantages over other methods. Most obvious is that the suction pipette allows regulations of the ionic interior of the cell. In addition the suction pipette offers a relatively low-resistance pathway to the cell interior that may not be achieved when, for example, a two-microelectrode clamp is used (see the chapter by Finkel and Gage). This allows one to apply relatively large currents to a cell without saturating electronic components with conventional voltage swings (±12 V); it may also increase clamp speed. Katz and Schwartz (14) have analyzed how the access resistance of the current-passing electrode may lead to a time constant that will slow a two-electrode voltage clamp. The combined clamp discussed in this chapter seems to have exceptional speed and stability characteristics at gains that yield negligible clamping errors. Such characteristics are not usually achieved with single-electrode clamping (see the chapter by Finkel and Redman). An additional advantage may be that the combined suction-pipette microelectrode may offer better spatial voltage control than a single-electrode clamp. For a single-electrode clamp the voltage is sensed from a point in a region of high current density, which may give a voltage gradient in the cell interior (8).

An additional advantage of the suction pipette is that its perfusion characteristics allow one to test properties of the electrophysiology of the intact cell (e.g., prevention of washout of Ca currents) (17). Such studies may not be possible with other methods of accessing the interior cell membrane such as the ripped-off inside-out patch technique.

We have studied fairly extensively the combined voltage-clamp system as it is applied to the study of Ca tail currents in snail neurons. We found that the bulk interior of the cell was spatially clamped within 3 mV at ~30 μs during a Ca tail current elicited by the cessation of a voltage-clamp step that fully activates Ca current. The series-resistance component was found to be some value <5 kΩ, and this should give rise to voltage-clamping errors due to ion conduction on the order of only a millivolt even during the largest Ca tail currents. Additionally capacitive current transients die out quite quickly; the fast component appears totally complete in 60 μs. Thus the capacitive current transient should lead to negligible series-resistance errors soon after voltage transition. Therefore we conclude that this combination of voltage-clamp system and preparation yields a viable method of studying relatively large, fast ionic currents.

Summary

The suction-pipette method of voltage clamp and internal perfusion is discussed. Included are fabrication details and design considerations for making suction pipettes for cells as small as mammalian ventricular myocytes with diameters of 10 μm or as large as snail neurons with diameters of 50–100 μm.

Exchange of intracellular ions as it relates to separation of membrane currents is described. Details of voltage control using a simple suction pipette or a combined suction pipette and microelectrode clamp are dealt with as well. Various electrical characteristics of the combined voltage clamp as applied to the snail neuron preparation were measured. Excellent fast spatial control was demonstrated by measurements with a second microelectrode. Series resistance was measured with a frequency-domain pseudorandom noise technique and was found to have negligible errors. Even large fast Ca tail currents, which would be expected to stress the clamp severely, are measured accurately with this method.

REFERENCES

1. Baker, P. F., A. L. Hodgkin, and T. I. Shaw. Replacement of the protoplasm of a giant nerve fibre with artificial solutions. *Nature London* 274: 379–382, 1961.
2. Brown, A. M., K. S. Lee, and T. Powell. Voltage clamp and internal perfusion of single rat heart muscle cells. *J. Physiol. London* 318: 455–478, 1981.
3. Brown, A. M., K. Morimoto, Y. Tsuda, and D. L. Wilson. Calcium current-dependent inactivation of calcium channels in *Helix aspersa*. *J. Physiol. London* 320: 193–218, 1981.
4. Brown, A. M., Y. Tsuda, and D. L. Wilson. A description of activation and conduction in calcium channels based on tail and turn-on current measurements. *J. Physiol. London* 344: 549–584, 1983.
5. Byerly, L., and S. Hagiwara. Calcium currents in internally perfused nerve cell bodies of *Limnea stagnalis*. *J. Physiol. London* 322: 503–528, 1982.
6. Byerly, L., and W. J. Moody. Intracellular Ca^{2+} and Ca currents in internally perfused snail neurons (Abstract). *Biophys. J.* 41: 292a, 1983.
7. Cole, K. S. Electrical properties of the squid axon sheath. *Biophys. J.* 16: 137–142, 1976.
8. Eisenberg, R. S., and E. Engel. The spatial variation of membrane potential near a small source of current in a spherical cell. *J. Gen. Physiol.* 55: 736–757, 1970.
9. Fenwick, E. M., A. Marty, and E. Neher. Sodium and calcium channels in bovine chromaffin cells. *J. Physiol. London* 331: 559–636, 1982.
10. Fishman, H. M. *Techniques in Cellular Physiology*. Amsterdam: Elsevier North-Holland, 1982, p. 1–50.
11. Fishman, H. M., L. E. Moore, and D. Poussart. *The Biophysical Approach to Excitable Membranes*. New York: Plenum, 1981, p. 65–87.
12. Fishman, H. M., D. Poussart, and L. E. Moore. Complex admittance of Na conduction in squid axon. *J. Membr. Biol.* 50: 43–63, 1979.
13. Hammil, O. P., A. Marty, E. Neher, B. Sakmann, and F. J. Sigworth. Improved patch-clamp techniques for high resolution current recording from cells and cell-free membrane patches. *Pfluegers Arch.* 391: 85–100, 1981.
14. Katz, G. M., and T. L. Schwartz. Temporal control of voltage clamped membranes: an examination of principles. *J. Membr. Biol.* 17: 275–291, 1974.
15. Kostyuk, P. G., O. A. Krishtal, and V. I. Pidoplichko. Intracellular dialysis of nerve cells: effect of intracellular fluoride and phosphate on inward current. *Nature London* 257: 691–693, 1975.
16. Kostyuk, P. G., O. A. Krishtal, and V. I. Pidoplichko. Calcium inward current and related charge movements in the membrane of snail neurons. *J. Physiol. London* 310: 403–421, 1981.
17. Kostyuk, P. G., N. S. Veselonsky, and A. Y. Tsyndrenko. Ionic currents in the somatic membrane of rat dorsal root ganglion neurons. II. Calcium currents. *Neuroscience* 6: 2431–2438, 1981.

18. Krishtal, O. A., V. I. Pidoplichko, and Y. A. Shakhovalov. Conductance of the calcium channel in the membrane of snail neurones. *J. Physiol. London* 310: 123-434, 1981.
19. Lee, K. S., N. Akaike, and A. M. Brown. Properties of internally perfused, voltage-clamped isolated nerve cell bodies. *J. Gen. Physiol.* 71: 489-507, 1978.
20. Lee, K. S., N. Akaike, and A. M. Brown. The suction pipette method for internal perfusion and voltage clamp of small excitable cells. *J. Neurosci. Methods* 2: 51-78, 1980.
21. Oikawa, T., G. S. Spyropoulos, I. Tasaki, and T. Teorell. Methods for perfusion of giant axon of *Liligo pealii*. *Acta Physiol. Scand.* 51: 195-293, 1961.
22. Takahashi, K., and M. Yoshii. Effects of internal free calcium upon the sodium and calcium channels in the tunicate egg analyzed by the internal perfusion technique. *J. Physiol. London* 279: 519-549, 1978.

EIGHT

Microelectrode Voltage Clamp: The Cardiac Purkinje Fiber

Robert S. Kass
Paul B. Bennett
University of Rochester School of Medicine and Dentistry, Department of Physiology, Rochester, New York

Theoretical Framework—Voltage Clamping Multicellular Tissue: Short cable: two-microelectrode voltage-clamp technique, Accuracy and limitations, Optimizing electrode spacings, Time-dependent changes for linear and nonlinear conductances, Summary • **Experimental Guidelines—Two-Microelectrode Technique:** Clamp circuit, Optimizing current microelectrodes, Clamp circuit modifications, Testing the clamp, Summary

Cardiac muscle, like skeletal muscle and nerves, is electrically excitable. Local stimulation by a brief electrical shock of adequate strength generates an impulse that propagates in a regenerative manner to distal regions. Although the diversity of electrical activity in the heart contrasts with the homogeneity of nerve and skeletal muscle action potentials, cardiac electrical impulses are generated by membrane permeability changes that generally resemble those in other excitable cells.

The mathematical framework used to describe impulse conduction in the heart is the same as that developed for nerves. It is based on the flow of local circuit currents (8) and provides the theoretical foundation for methods of measuring membrane current in cardiac cells.

In this chapter the application of a microelectrode voltage-clamp technique that has been used to characterize membrane currents in the mammalian cardiac Purkinje fiber is discussed. The chapter is divided into two sections. The first section describes the theoretical background for this technique in the Purkinje fiber. The second section is an experimental guide to applying this technique in the laboratory.

Theoretical Framework
Voltage Clamping Multicellular Tissue

In mammalian hearts, impulses are rapidly conducted throughout both ventricular chambers via a system of specialized conducting tissue referred to as Purkinje fibers [see Kass and Scheuer (11) and Tsien and Siegelbaum (18) for review]. In many species (e.g., rabbit, dog, calf, and sheep) these fibers appear as free-running strands on the endocardial surfaces and may be easily dissected from the heart.

Voltage spread in isolated Purkinje fibers is well-described by the linear one-dimensional cable equation (10)

$$\frac{a}{2R_a} \frac{\partial^2 V}{\partial X^2} = I_m = I_i + C_m \frac{\partial V}{\partial t} \qquad (1)$$

where a is fiber radius, X is distance, R_a is myoplasmic resistivity, V is membrane potential, I_m is membrane current, I_i is ionic current, C_m is membrane capacity per unit area, and t is time. A powerful simplification of this relationship occurs experimentally when the membrane potential is uniformly controlled along the cable by passing current from an intracellular source. This condition is referred to as voltage clamp. In this case the capacity current (except for brief transients) is eliminated, and the measurable applied current equals membrane current. Ideally this current source will equally affect all areas of active membrane to avoid longitudinal voltage gradients. This criterion can be satisfied in nerve by inserting a pair of axial wires along the interior of the squid giant axon. One wire monitors intracellular potential, and the other passes current to control the transmembrane potential. This procedure provides uniform voltage control along the length of the axon.

It is not possible to insert an axial wire to short-circuit intracellular resistance in isolated Purkinje fibers because these preparations consist of bundles of much smaller individual cells. Purkinje fibers are columns of closely packed cells that are electrically well coupled by specialized gap junctions. Each cell is ~10 µm in diameter and 100 µm in length, and the entire bundle is encased in a tough layer of connective tissue. Because space-clamp conditions cannot exist in this preparation, voltage-clamp techniques must be carried out under conditions in which spatial and temporal voltage nonuniformity exist but are kept within reasonable limits of error. The nature of these limitations must be understood before undertaking voltage-clamp experiments in the Purkinje fiber or any other multicellular preparation.

Short cable: two-microelectrode voltage-clamp technique

Until very recently all voltage-clamp experiments in cardiac tissues were carried out in isolated cell bundles. Two experimental arrangements, both subject to restrictions of voltage nonuniformity, have been used: the two-microelectrode voltage clamp and the sucrose-gap voltage clamp (for review see ref. 7). The discussion here focuses on the former technique because it has been used successfully in the Purkinje fiber.

Figure 1 illustrates the two microelectrode technique as it was first introduced by Deck et al. (5) and described by Weidmann (20). This technique relies on the fact that there is normally good electrical coupling between the cells within a Purkinje fiber. If the fiber is cut into short segments, the cut ends of the fiber "heal over" and the preparation then behaves electrically as a short cable terminated by very high resistances. Thus there is little or no longitudinal current flow across the surface membranes of the cells at the cut ends (3, 19).

Deck et al. (5) were aware of this healing-over property of the cardiac multicellular preparation and reasoned that by making fibers short enough they could minimize voltage gradients in the resulting short cable, thus making a preparation that could be voltage clamped with a reasonable amount of confidence. They proposed that membrane potential, measured via an intracellular microelectrode at one point along the

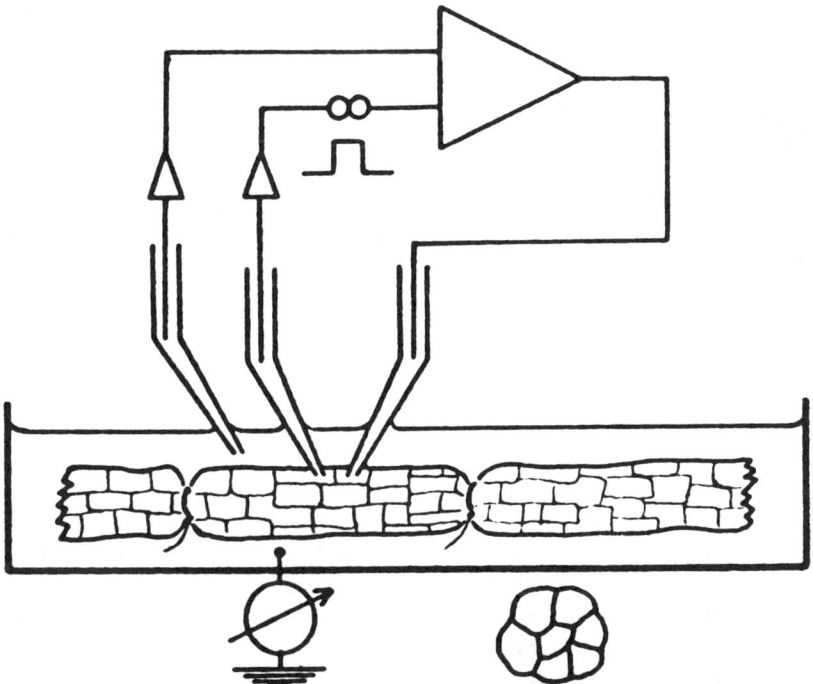

Fig. 1. Two-microelectrode voltage-clamp arrangement for the cardiac Purkinje fiber. Shortened segments of Purkinje fiber cell bundles are obtained by cutting or ligating. Cross section of resulting short fiber (*lower right*) shows presence of several cells. Intracellular voltage is measured at 1 location and compared with command pulse at summing junction of operational amplifier. Voltage control at this point is provided by current passed through 2nd microelectrode impaled at another location in fiber. Current crosses cell membrane and is collected and measured by virtual ground amplifier. [From Weidmann (20)].

cable, can be controlled by passing current through a second microelectrode at another location (Fig 1). Provided the length of the cable is sufficiently short (<1 space constant), the membrane potential measured by a microelectrode at a single location is a reasonable estimate of the average potential along the cable. Initial experiments with a second voltage-measuring microelectrode showed that voltage control was adequate under certain conditions (5). However, several questions about quantitative restrictions of the technique have only recently been addressed (13).

For example, what are the optimum electrode spacings for this procedure? Can this approach be used to measure inward currents? Are there differences between electrode spacings for measurement of inward and outward currents? Can this technique be empirically tested? These questions are addressed in the next section.

Accuracy and limitations

To test the accuracy of microelectrode voltage clamping in the Purkinje fiber, an experimental procedure was developed to evaluate the measured current. One approach to this problem was provided by Kass et al. (13), who demonstrated that comparison of two independent measurements of membrane current could serve as a method for checking the validity of both techniques. They used a two-microelectrode procedure as one estimate of membrane current. In this procedure (Fig. 2B), the total injected current (I_T) is used as a measure of transmembrane current flowing at a representative membrane potential (V_1). For the second measurement they adapted the three-microelectrode technique of Adrian et al. (1) and referred to this measurement of membrane current as $I_{\Delta V}$. The strategy used by Kass et al. (13) to evaluate these procedures is outlined in this section.

Figure 2 shows the electrode arrangement of Adrian et al. (1) and its modification for studies of the Purkinje fiber. In each case one end of the fiber is indicated in the figure ($x = 0$), and two voltage-recording electrodes (V_1 and V_2) are shown impaling the preparations at separate sites. In the case of the Purkinje fiber (Fig 2B), the other end of the fiber is also indicated ($x = 2h$, where h is the hemilength of the fiber). Current (I_T) is passed via another microelectrode to control the membrane potential measured at V_1. Current passed via this electrode eventually leaves the fiber by crossing the surface membrane along the entire length of the cable. In the two-microelectrode technique for the short cable, this current is also used as a measure of total transmembrane current. The voltage measured at V_1 is then taken as the average membrane potential at which this total current flows.

The three-electrode technique takes advantage of the fact that trans-

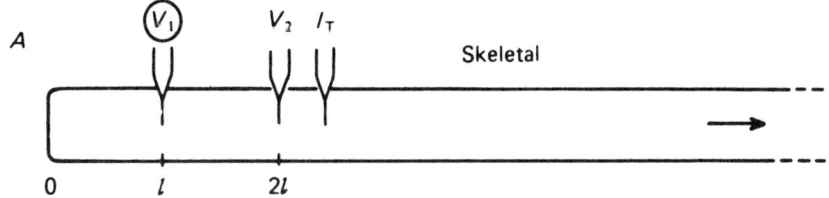

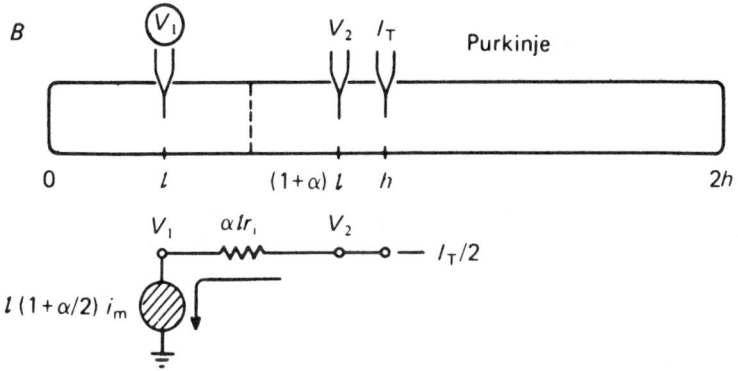

Fig. 2. Three-microelectrode voltage clamp. *A*: method originally used in skeletal muscle (1). Voltage control is imposed at V_1 and membrane current density is reported by $V_2 - V_1$. Total applied current (I_T) is not easily interpreted because of unequal current flow into long and short ends of fiber. *B*: technique adapted for regenerative inward currents in Purkinje fibers. Current-passing electrode is located at distance h from each sealed end so I_T can be interpreted. Spacing (αl) between V_2 and V_1 is adjusted to improve accuracy of $V_2 - V_1$ for negative membrane conductance. Optimal spacing for linear positive-slope conductance case ($\alpha = 1.5$) is shown. Signal ($V_2 - V_1$) approximates I_m (membrane current density) over region from sealed end to *dashed line*. *Below*: transmembrane current flow in the lumped equivalent circuit (*shaded element*). [From Kass et al. (11).]

membrane current near the end of the cable is supplied by longitudinal current that originates at the current-passing micropipette. This longitudinal current flow, which exists because current is passed from a point source into a distributed cable, produces an ohmic (IR) voltage drop across the internal resistance of the fiber. The voltage drop is measured as the difference between V_2 and V_1 ($V_2 - V_1 = \Delta V$). Adrian et al. (1) showed that ΔV is proportional to transmembrane current density measured near the V_1 electrode. Kass et al. (13) showed that, in the short cable, I_T and ΔV can be used as simultaneous measures of membrane current. Therefore the theory was used to design experimental tests of membrane current using both the ΔV and I_T measurements of membrane current.

Optimizing electrode spacings

In the two-microelectrode technique, which uses I_T as a measure of membrane current, the current-passing electrode should be placed midway between the cut ends of the preparation. This provides symmetrical current flow into both fiber halves. However, optimum location of the voltage electrode(s) in either the two- or three-microelectrode technique is not obvious. Kass et al. (13) showed that the accuracy of current measurement is very dependent on electrode spacing.

Following the precedent of Adrian et al. (1) and Schneider and Chandler (16), Kass et al. (13) derived an expression for current density near the V_1 electrode that is given as

$$I_m = p \frac{1}{\alpha(1 + \alpha/2)l^2 r_i} \Delta V \qquad (2)$$

where l is the distance between V_1 and the cut end, αl is the interelectrode (V_1 and V_2) spacing, r_i is the intracellular resistance, and p is an arbitrary correction factor. The correction factor p is an index of the accuracy of the technique. When $p = 1$, ΔV exactly equals membrane current density at the V_1 recording site. This measure of membrane current is less accurate when p differs from unity.

Figure 3 shows that variation in electrode spacing influences the correction factor. The figure shows the results of computations for two situations in which closed-form analytical solutions exist for the cable equations. The first case is that of a linear cable with a passive positive-slope membrane conductance ($g_m > 0$). In this case the voltage distribution varies as a hyperbolic cosine (cosh) function (10, 19). The expression for p is then given by

$$p = \frac{1}{2} \frac{(\alpha + 2)l^2}{\lambda^2} \frac{\cosh(l/\lambda)}{\cosh[(\alpha + 1)l/\lambda] - \cosh(l/\lambda)} \qquad (3)$$

where λ is the space constant. For this situation the symmetric electrode spacing ($\alpha = 1$) results in the widest range of membrane conductances that can be measured with errors <15% (Fig. 3). However, this is not the best electrode arrangement for the measurement of inward current.

The simplest model used to simulate a membrane with inward current is the linear cable with a constant negative-slope membrane conductance ($g_m < 0$). Now the solution for the short-cable equation is given by a cosine function, and the expression for the correction factor changes to

$$p = \frac{1}{2} \frac{\alpha(\alpha + 2)l^2}{\lambda^2} \frac{\cos(l/\lambda)}{\cos[(\alpha + 1)l/\lambda] - \cos(l/\lambda)} \qquad (4)$$

When this expression is evaluated for different conductances and elec-

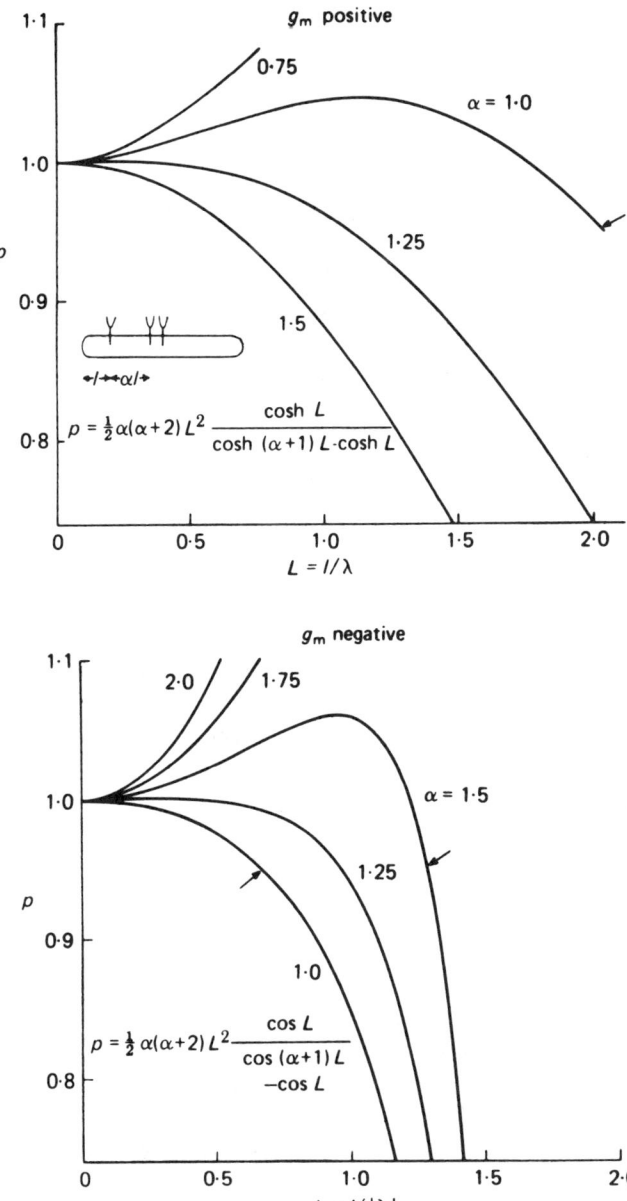

Fig. 3. Correction factor p as function of electrode spacings. Abscissa gives electrotonic distance from V_1 to sealed end. Parameter α is spacing ratio (*inset*). *Arrows* mark point at which p falls to 0.95; g_m, membrane conductance.

trode spacings, a new optimum arrangement is determined. The calculations showed that the range of measurable negative-slope conductances can be extended by changing to an unsymmetrical electrode arrangement ($\alpha = 1.5$) (Fig. 3B). According to the computations, rearrangement of the electrode spacing quadruples the maximum negative-slope conductance that can be measured within a 15% error spread.

Thus careful electrode arrangement improves the accuracy of the ΔV approximation of membrane current density, and the optimum arrangement depends on the nature of the current under investigation. However, with improved electrode spacing, can the two-microelectrode technique be used to monitor membrane current encountered in the Purkinje fiber? This question is addressed in the next section.

Time-dependent changes for linear and nonlinear conductances

In any voltage-clamp experiment a command voltage and a measurment of membrane potential are compared at the input of a feedback amplifier. This comparison is used to control membrane potential. When the voltage-clamp procedure is applied to a short cable, voltage is measured at one site along the cable and current is injected at a different site (see Fig. 1). This displacement in distance between the voltage sensor and current injector results in a lag in the response time of the desired voltage change (at the voltage-recording site). It is important to understand the origin of this phase lag and to recognize the differences that result when studying linear or nonlinear conductances.

Linear conductances. Consider the progression of events that occurs when voltage clamping a short cable in which the membrane conductance consists of a linear (non-voltage-dependent) leak conductance (g_l)

$$I_m = g_l V_m \tag{5}$$

where g_l is a constant and V_m is membrane potential. This situation corresponds to a voltage-clamp experiment in which all of the time- and voltage-dependent conductances have been blocked.

Figure 4 shows the events that occur when this model cable is voltage clamped. Voltage is measured at one point (V_1) and then compared with the command potential (V_c) at the summing junction of a feedback amplifier. The total injected current is determined by the gain of this feedback amplifier

$$I_T = \text{gain}(V_c - V_1) \tag{6}$$

Figure 4A shows the currents that result when the command voltage is stepped from rest (0 mV) to a test potential of 42.5 mV. The clamp step has been rounded with a 1-ms time constant, and the voltage recorded at V_1 is also shown.

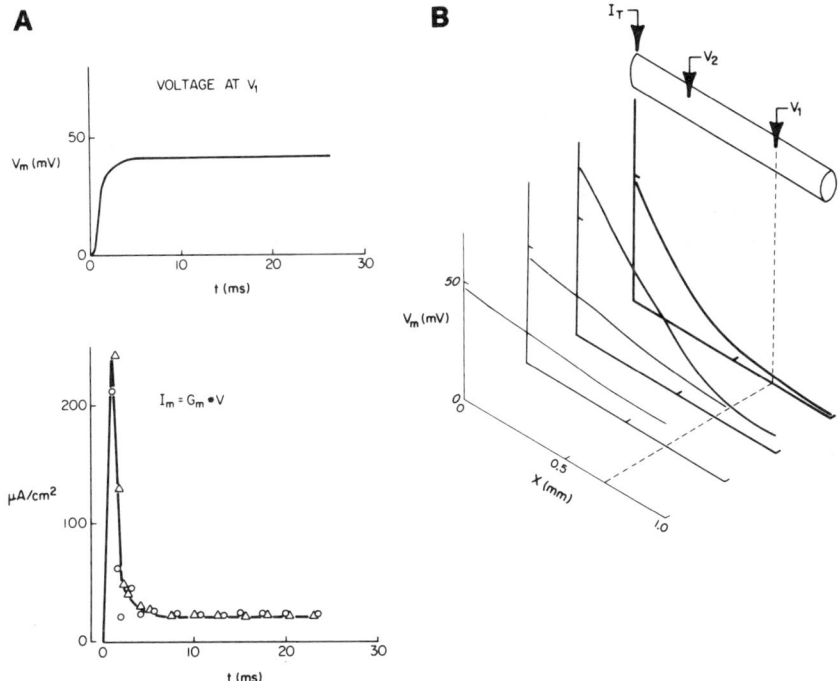

Fig. 4. Computational simulation of voltage clamp of linear cable. Voltage distribution in linear cable depends on distance from current injector and time. *A*: voltage clamp step (*upper panel*) recorded at electrode V_1 after command pulse from 0 mV to potential of 42.5 mV. *Lower panel* illustrates resulting currents as function of time during clamp step. *Curve*, computed membrane current at V_1; ○, current measured using I_T; △, ΔV measurement of membrane current. In this simulation, command pulse was filtered with time constant of 1 ms, and nonlinear conductance properties of cable were set to zero so that only ohmic conductances are present. *B*: three-dimensional illustration of spacial and temporal distribution of voltages along cable. Relative distances and electrode spacings illustrated by cable. Voltage distributions along cable are shown (*back to front*) 0.5, 1, 2.0, and 20 ms, respectively, after clamp step is imposed. Membrane time constant (τ_m) is 14 ms, resting space constant (λ) is 2 mm.

The total current (open circles) displays a large rapid outward transient that corresponds to the charge transfer needed to change the voltage on the membrane capacitance. The panel shows three different estimates of this current. The points labeled I_T were determined from Equation 6. The smooth curve is the membrane current (I_m) computed at V_1 using Equation 5. The points labeled ΔV were determined by the voltage difference ($V_2 - V_1$) along the cable at different times during the voltage-clamp step. Comparison of the two experimental measures of membrane current (I_T and ΔV) to the known membrane current (I_m) allows evaluation of the accuracy of these two experimental techniques. For this conductance both I_T and ΔV agree well with the actual (or theoretical) membrane current (Fig. 4).

What happens to the distribution of voltages along this cable as V_1 is stepped to a new value? Figure 4B shows instantaneous voltage distributions along the cable at several different times during the voltage step.

Initially the fiber is at rest, and all voltages along its length are zero. When the positive clamp step is first imposed, V_1 is still zero and a large difference exists between V_1 and V_c. This results in a large surge of injected current (I_T) that establishes a voltage gradient between the current-passing electrode and the V_1 electrode. As the cable becomes charged, V_1 becomes more positive; the difference between V_1 and V_c is reduced, and the voltage profile along the cable progressively becomes less steep (see $t = 0.5$ ms through $t = 20$ ms). Finally the steady-state voltage profile shows a small decrement from the current electrode to the end of the fiber. Thus a finite difference exists between V_c and V_1, and steady current must be passed to minimize this difference. The resulting voltage profile is described by the well-known steady-state solution for the linear short cable (see *Optimizing electrode spacing*, p. 176).

From the voltage profiles in Figure 4B, it is reasonable to expect that the experimental estimates of membrane current are least accurate when the voltage distributions show the steepest gradients (times briefer than 2 ms) and this suspicion is confirmed in the current traces in Figure 4A.

Nonlinear conductances: inward current. Consider what happens when we voltage clamp a cable in which the membrane conductance is nonlinear, as for the cardiac Purkinje fiber. In this case the membrane conductance is assumed to consist of a linear leak plus a time-dependent inward current. To simplify the computation, we assumed that membrane current is described as

$$I_m = g_l V_m + \bar{g}_{Ca^{2+}}\, d(t) f(t)(V_m - V_{Ca^{2+}}) \tag{7}$$

This approximates the situation in which Ca channel current is being investigated. The term $\bar{g}_{Ca^{2+}}$ is the maximum Ca channel conductance, g_l is a constant, V_m is the membrane potential, and $V_{Ca^{2+}}$ is the Ca equilibrium potential. The time-dependent gating variables are estimated by $d(t)$ (activation) and $f(t)$ (inactivation). The time-dependent parameters for this computation (see legend of Fig. 5) were chosen to provide insight into the changes that develop along voltage-clamped cables and are thus slower than comparable parameters reported for Ca channel current in cardiac cells (9).

Figure 5 shows the results of the computations. Again the fiber is initially at rest; i.e., the membrane potential is 0 mV along the entire cable at $t = 0$. When the clamp step is first imposed, the current and voltages (Fig. 5B) change in a manner very similar to the previous computation for a linear conductance because at early times ($t < 1$ ms) the inward Ca conductance has not yet turned on and the membrane conductance is in fact linear. However, as time progresses and Ca current begins to contribute to membrane current, the voltage profile along the cable changes in a very interesting manner.

At $t = 1$ ms, the membrane current approaches zero because $I_{Ca^{2+}}$ almost balances $g_l V_m$ (leak current). At this time the cable is almost uniformly charged (Fig. 5B, $t = 1$ ms). But as membrane current becomes more inward, the voltage profile begins to reverse—voltages become more positive at V_1 than they are at the current electrode. This reversed voltage gradient is the result of the nonlinear inward membrane current that adds positive charge to the inner surface of the membrane capacitance and depolarizes the membrane. The gradient reverses the sign of

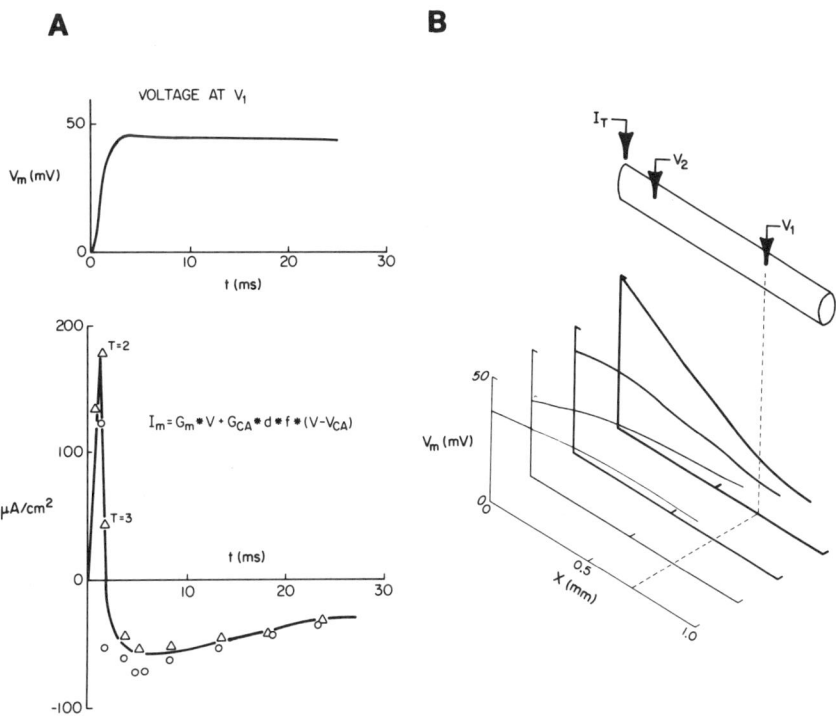

Fig. 5. Computer simulation of voltage clamp of nonlinear cable with time-dependent inward current. Conditions similar to those in Fig. 4 except that additional inward current component is incorporated into model. Total membrane current was generated by Eq. 7 (where $g_{Ca^{2+}} = 2$ mmhos/cm^2 and $\tau_f = 80$ ms) (see Fig. 3, ref. 12). Voltage distributions along cable are shown (*back* to *front*) 0.5, 1.0, 4.5, and 20 ms, respectively, after voltage step is imposed.

the ΔV and I_T measurements and accounts for the inward current measured under these conditions.

The steady-state voltage profile for this nonlinear conductance (see Fig. 3B of ref. 11) clearly contrasts with the profile for a linear conductance (see Fig. 4). In Figure 5 there is an 8-mV voltage difference between the current electrode and the end of the fiber, and this voltage gradient becomes more severe for larger nonlinear conductances. For an infinite cable the voltage would become progressively more positive as distance from the current source increased. This can lead to propagation of all-or-none responses in distal ends of these longer preparations (see Fig. 8 of ref. 1). The short cable circumvents this problem—the size of the voltage gradient along the fiber is limited by fiber length.

The voltage profiles in Figure 5B suggest that the greatest error in experimental measurement of current occurs at the peak of the inward current ($t = 4.5$ ms); this is clear in the computed membrane currents in Figure 5A. The computation shows that for the maximum inward current in this example both ΔV and I_T are good estimates of membrane current. But it also suggests that these estimates break down when peak inward current is made sufficiently large.

Kass et al. (11) tested the accuracy of ΔV and I_T as functions of peak inward current studied for steady-state voltage profiles. They found that both experimental measures were quite reasonable, provided that peak inward current is <100 $\mu A/cm^2$. When peak current is kept within this limit, the computations indicate that even I_T (which is always less accurate than ΔV) reflects membrane current to within 15% error (Fig. 6).

These computations also showed that comparison of ΔV and I_T measurements can provide an empirical check of the voltage-clamp data. That is, when I_T and ΔV agree (Fig. 6), they both agree with the actual membrane current density. Discrepancy between the two measurements indicates that I_T is more likely to differ from the actual membrane current but that neither measurement should be trusted. This result is independent of the cable geometry or the axial resistance chosen. Thus, although this test has been applied in cardiac preparations, this procedure can be used to test the validity of membrane currents in other multicellular preparations (e.g., smooth muscle strips) as well.

Summary

In summary, theoretical computations have been used to determine voltage distributions in model cables for linear and nonlinear membrane conductances. The calculations reveal an unusual voltage profile for nonlinear conductances, which produce voltage gradients that can lead to propagation of all-or-none responses in long preparations. Theory

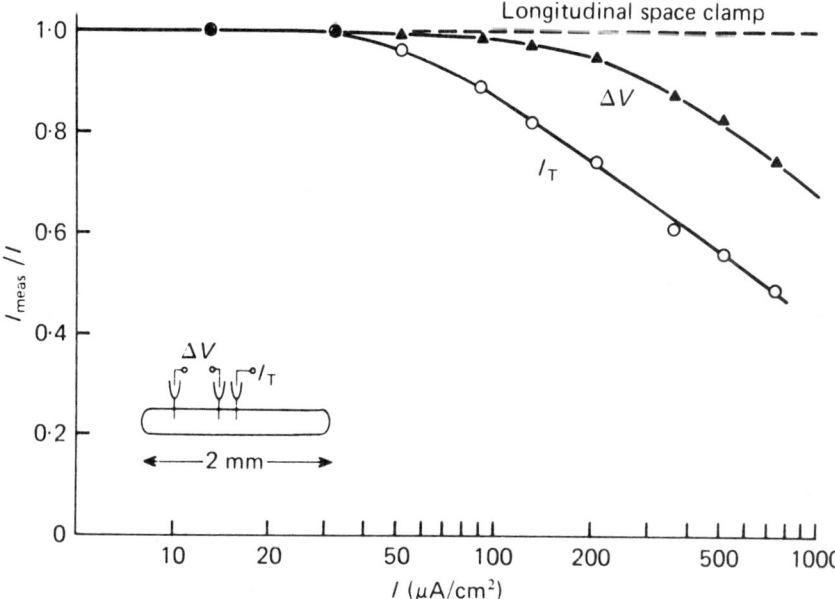

Fig. 6. Accuracy of ΔV and I_T as function of inward-current density. Ordinate is I_{meas} (current density measured by ΔV or I_T) divided by I (genuine value of membrane current). Values of ΔV, I_T, and I are given at V_1. Quotient of unity for all I (*dashed line*) corresponds to ideal case of longitudinal space clamp.

predicts that shortened preparations and optimum electrode spacing can significantly extend the range of measurable membrane conductances by both the three- and two-microelectrode voltage-clamp procedures. In short fibers the current-passing electrode should be placed midway between the ends of the preparations. For the measurement of inward current the voltage-measuring electrode (V_1) should be placed two-thirds the distance between the current electrode and the cut end. For the measurement of outward currents the voltage electrode should be placed one-half the distance between the current source and the end of the fiber (also see ref. 16).

Comparison of these two measures of membrane currents provides an empirical test of the validity of these measurements. Such comparisons have been used experimentally to test the accuracy of voltage-clamp measurements of Ca current in the cardiac Purkinje fiber (13) and ventricular muscle (15) and of Na current in the rabbit Purkinje fiber (2, 3). These studies have confirmed the accuracy of the two- and three-microelectrode procedures in these preparations. Although both techniques are acceptable, the two-microelectrode approach is the procedure of choice because of its greater experimental simplicity.

Experimental Guidelines—Two-Microelectrode Technique

Clamp circuit

The experimental arrangement for the two-microelectrode technique remains very similar to that described originally by Deck et al. (5). The basic circuit (Fig. 7) is very simple and should not require a commercially wired component. As shown in Figure 7 the membrane potential measured at recording site V_1 (see Fig. 1) is compared with a command signal (V_c) at the summing junction of an operational amplifier. Current is passed via the current microelectrode (I_T) to maintain the difference $V_c - V_1$ close to zero.

Optimizing current microelectrodes

Probably the greatest cause of experimental problems in this technique is the current source: the current-passing micropipette. The current-passing characteristics of microelectrodes with small tip diameters are highly nonlinear and very dependent on the electrolyte in the pipette. Failure to uniformly inject adequate current results in the inability to control the voltage at V_1, regardless of the nature of the ionic current under investigation.

Glass microelectrodes suitable for impaling cardiac cells typically have tip resistances of ~10–20 MΩ when they are filled with 3 M KCl. But this electrolyte is not very well suited for passing current because these electrodes show rectification. On the other hand, when the same pipettes are filled with 1.5 M potassium citrate, they show less rectification but are characterized by four to five times higher resistances. Furthermore in some cases other electrolytes that result in even higher resistances may be necessary (12). Because of these large electrode resistances, the feedback amplifier in Figure 6 must be able to produce large voltage changes at the output stage to pass adequate current through the micropipette. Thus the first special requirement of this arrangement is that this amplifier be a high-voltage operational amplifier (note, these amplifiers require high-voltage power supplies). A good amplifier and companion power supply are suggested (Fig. 7).

Special attention should be paid to the construction of the current electrodes because they are so critical to these experiments. Two very useful and simple steps can be taken to optimize these electrodes: choice of pipette glass and electrode beveling. First, electrodes used for current passing should be pulled from thin-wall electrode glass (typical dimensions are 1.2-mm outer diam, 0.2-mm wall thickness). [Many suppliers stock such glass, but two companies in particular that offer thin-wall

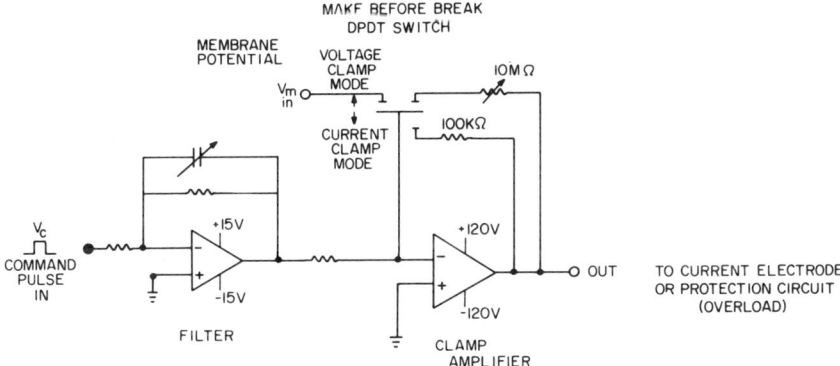

Fig. 7. Simplified diagram of voltage-clamp circuit. Command pulses of appropriate magnitude and duration are input to inverting operational amplifier (TI UA741CP) configured as filter to attenuate high-frequency components of command pulse (rounding). Inverted pulse then enters summing junction of high-voltage operational amplifier (Teledyne-Philbrick 1032 FET), and switch is used to configure circuit for voltage clamp or current clamp. Teledyne-Philbrick 2217, 120-volt power supply is used to provide power for high-voltage operational amplifier.

(Omega-dot) glass are World Precision Instruments (WPI), Inc., West Haven, CT, and Glass Company of America, Bargaintown, NJ.]

Second, beveling of the electrode tip greatly improves its current-passing characteristics. Several simple satisfactory electrode-beveling procedures are described in the literature (4, 14). These arrangements can be modified to create very inexpensive beveling systems. For example, in our laboratory 60-cycle noise of a grounded microelectrode is monitored as the tip is beveled. This noise appears to be roughly linearly related to the tip resistance and thus provides a simple continuous monitor of changes in the tip. The electrode is beveled by stirring a solution containing abrasive silicon carbide (grit #120) powder (Buehler, Ltd., Evanston, IL) with a magnetic stir bar. Electrodes can be beveled to selected tip resistances in this manner. The speed and degree of beveling can be chosen by varying the velocity of the stir bar and the concentration of the abrasive powder (e.g., ref. 4).

Because the tip resistances of these microelectrodes are inherently high, any stray capacitance in the voltage-clamp circuit slows the response time of the clamp. Capacity coupling exists between the voltage-recording and current-passing micropipettes in the two-electrode voltage clamp, and this capacitance must be minimized. The simplest approach to this problem is to lower an insulated grounded conducting sheet (thin brass insulated with a lacquer such as nail polish) between these electrodes. This will reduce but not eliminate the capacitance. A more effective approach is to coat each microelectrode with an insulated conductive layer (see the chapter by Sachs).

Clamp circuit modifications

Overload protection. Tsien (17) has described a very useful circuit (Fig. 8) that prevents accidental passage of very large amounts of current through the current electrode. This safeguard circuit interrupts the current output of the feedback amplifier when the clamp output voltage exceeds ±100 V. Such a circuit greatly increases the chances of carrying out successful experiments and is highly recommended.

Rounding command pulse: ringing in clamp. Another important modification of the voltage-clamp circuit is necessary because of the response time of the voltage at the V_1 recording site, which is caused by the cable properties of the preparation being studied (see *Time-dependent changes for linear and nonlinear conductances*, p. 178). This delay in the voltage response introduces a phase lag between the measured voltage and the desired command potential. The command pulse can be rounded by a resistance-capacitance (*RC*) circuit to minimize effects of this phase lag. Without proper rounding the voltage-clamp circuit can oscillate and distort measured current. Such oscillations are referred to as *ringing* in the clamp circuit. A clear understanding of the causes, detection, and prevention of clamp ringing is very important before using the voltage-clamp technique.

Figure 9 illustrates the influence of command-pulse rounding on clamp ringing. In Figure 9A a square command pulse is imposed on the same

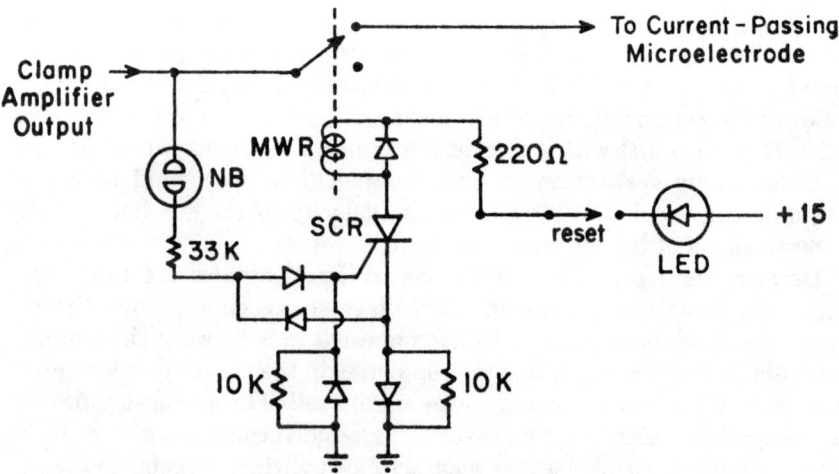

Fig. 8. Clamp overload protection circuit. [From Tsien (17)].

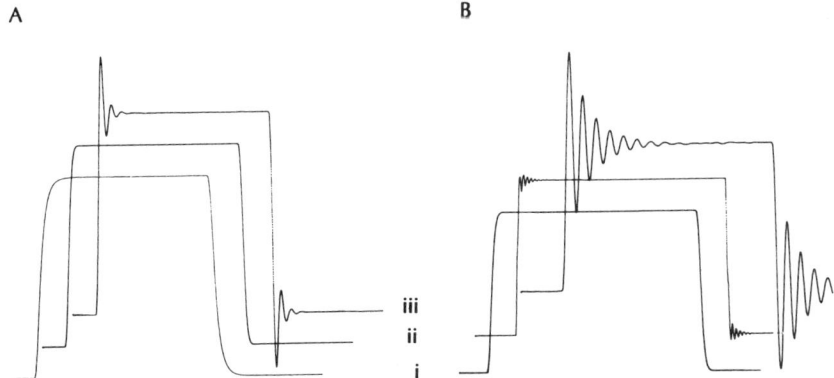

Fig. 9. A: influence of command pulse rounding on ringing in clamp circuit. *i*, Too much rounding; *ii*, optimum attenuation of high-frequency components in command pulse; *iii*, square-wave command pulse in which high-frequency components are not removed (no rounding). These results would be achieved in practice by altering capacitor in feedback loop of filter (Fig. 7). B: influence of voltage-clamp gain on ringing in clamp circuit. *i*, Optimum gain setting for clamp circuit to achieve rapid transition to test potential without ringing; *ii*, gain was increased until circuit began to become unstable; *iii*, additional small increases in gain produced large oscillations in clamp circuit. Oscillations should be avoided by monitoring current and voltage traces and by adjusting command-pulse rounding and clamp feedback gain. Results shown in both panels were generated using a theoretical cable with properties given in legend to Fig. 4.

cable that is illustrated in Figure 4. When V_c is not rounded, the voltage at V_1 overshoots the desired voltage and the feedback amplifier passes inward current to correct this problem (Fig. 9A, *trace iii*). This compensation produces damped oscillations in the applied current and in the measured voltage traces that settle as V_m approaches V_c.

In Figure 9A (*trace ii*) the command pulse has been rounded with an exponential time constant chosen to minimize the phase lag between V_c and V_1. Clearly the clamp does not ring—no oscillations are apparent in V_1 because V_1 can gradually approach V_c. Figure 9A (*trace i*) also shows the effects of too much rounding; the rise time of the voltage-clamp step is compromised.

Voltage-clamp stability is also affected by the gain of the feedback amplifier. Figure 9B illustrates the effects of gain on ringing in the voltage-clamp circuit. As the feedback gain is increased, oscillations appear in the voltage trace indicating that the circuit has become unstable (ringing). Thus the experimenter is faced with a trade off between being assured that the clamp gain is sufficiently high so that voltage control is rapidly achieved and maintained and being assured that the gain is not so high that the circuit becomes unstable and begins to ring. Computer modeling and actual experience with voltage clamp of Purkinje fibers have shown that the measured current is an even more sensitive indicator of instability in the clamp circuit. Experience indicates that voltage and total current should be monitored for ringing when the membrane conductance is nearly linear. Command-pulse rounding and clamp feedback gain should be adjusted to provide optimal response time of the clamp without detectable ringing in the current trace.

Testing the clamp

Before carrying out experiments on biological tissue, the voltage-clamp circuit should be tested with simple analog circuits. Two model circuits are very useful for such purposes. One circuit is an analog equivalent of a passive electrical cable with parameters that resemble those reported for the Purkinje fiber (19). With this circuit the current can be injected at one end of the cable from the output of the voltage-clamp circuit. Voltage may be measured at various locations along the cable and then fed back to the clamp-amplifier summing junction. Voltages at other locations may also be monitored to study voltage responses along the cable. Thus the analog circuit can be used to generate the same responses that were computed for Figures 4, 8, and 9, and the points discussed with these figures can be directly studied.

Finally, a simple RC circuit is very useful for testing the response time of the clamp and the effects of electrode shielding. This circuit should be placed between the tissue bath and the input to the current-to-voltage converter. In this manner the circuit behaves as a local patch of membrane, with the outer surface fixed at 0 mV (input-to-current amplifier) and the inner surface represented by the tissue bath. Then, if the current and voltage microelectrodes are positioned in the bath, the inner surface of this model membrane can be voltage clamped. Because microelectrodes are used, this model provides a good method of calibrating the clamp response time under conditions resembling those of actual experiments. In addition the test permits direct evaluation of the effectiveness of electrode shielding for this arrangement.

Summary

The material presented in this section is intended to serve as a guide to the application of the two-microelectrode voltage-clamp technique to multicellular cardiac preparations and other tissues. Several experimental pitfalls are pointed out, and some suggestions are presented to help avoid these problems. Nevertheless difficulties that have not been presented are bound to surface.

This chapter serves as a general guide for the use of the two-microelectrode voltage clamp; however, experience in a laboratory that uses this technique is probably the most efficient way to master this procedure.

REFERENCES

1. Adrian, R. H., W. K. Chandler, and A. L. Hodgkin. Voltage clamp experiments in striated muscle fibres. *J. Physiol. London* 208: 607–644, 1970.
2. Colatsky, T. J. Voltage clamp measurements of sodium channel properties in rabbit cardiac Purkinje fibres. *J. Physiol. London* 305: 215–234, 1980.
3. Colatsky, T. J., and R. W. Tsien. Sodium channels in rabbit cardiac Purkinje fibres. *Nature London* 278: 265–268, 1979.
4. Corson, D. W., S. Goodman, and A. Fein. An adaptation of the jet stream microelectrode beveler. *Science* 205: 1302, 1979.
5. Deck, K. A., R. Kern, and W. Trautwein. Voltage clamp technique in mammalian cardiac fibers. *Pfluegers Arch. Gesamte Physiol. Menschen Tiere* 280: 50–62, 1964.
6. DiFrancesco, D., and P. A. McNaughton. The effects of calcium on outward membrane currents in the cardiac Purkinje fibre. *J. Physiol. London* 289: 347–373, 1979.
7. Fozzard, H. A., and G. W. Beeler. The voltage clamp and cardiac electrophysiology. *Circ. Res.* 37: 403–413, 1975.
8. Hodgkin, A. L., and W. A. H. Rushton. The electrical constants of a crustacean nerve fiber. *Proc. R. Soc. London Ser. B* 133: 444–479, 1946.
9. Isenberg, G., and U. Klockner. Calcium currents of isolated bovine ventricular myocytes are fast and of large amplitude. *Pfluegers Arch.* 395: 30–41, 1982.
10. Jack, J. J. B., D. Noble, and R. W. Tsien. *Electrical Current Flow in Excitable Cells.* Oxford, UK: Oxford Univ. Press, 1975.
11. Kass, R. S., and T. Scheuer. Calcium ions and cardiac electrophysiology. In: *Calcium Blockers: Mechanisms of Action and Clinical Applications*, edited by S. F. Flaim and R. Zelis. Baltimore, MD: Urban and Schwarzenberg, 1982 p. 3–19.
12. Kass, R. S., T. Scheuer, and K. J. Malloy. Block of outward current in cardiac Purkinje fibers by injection of quaternary ammonium ions. *J. Gen. Physiol.* 79: 1041–1063, 1982.
13. Kass, R. S., S. A. Siegelbaum, and R. W. Tsien. Three-microelectrode voltage clamp experiments in calf cardiac Purkinje fibers: is slow inward current adequately measured? *J. Physiol. London* 290: 201–225, 1979.
14. Lederer, W. J., A. J. Spindler, and D. A. Eisner. Thick slurry beveling. A new technique for beveling extremely fine microelectrodes and micropipettes. *Pfluegers Arch.* 381: 287–288, 1979.
15. Reuter, H., and A. Scholz. A study of the ion selectivity and the kinetic properties of the calcium dependent slow inward current in mammalian cardiac muscle. *J. Physiol. London* 264: 17–47, 1977.
16. Schneider, M. F., and W. K. Chandler. Effects of membrane potential on the capacitance of skeletal muscle fibers. *J. Gen. Physiol.* 67: 125–163, 1976.
17. Tsien, R. W. Effects of epinephrine on the pacemaker current of cardiac Purkinje fibers. *J. Gen Physiol.* 64: 293–319, 1974.
18. Tsien, R. W., and S. Siegelbaum. Excitable tissues: the heart. In: *The Physiological Basis for Disorder of Biomembranes*, edited by T. Andreoli, J. F. Hoffman, and D. Fanestil. New York: Plenum, 1978, p. 517–538.
19. Weidmann, S. The electrical constants of Purkinje fibres. *J. Physiol. London* 118: 348–360, 1952.
20. Weidmann, S. Cardiac electrophysiology in the light of recent morphological findings. *Harvey Lect.* 61: 1–15, 1966.

NINE

Space-Clamp Problems When Voltage Clamping Branched Neurons With Intracellular Microelectrodes

Wilfrid Rall
Idan Segev

Mathematical Research Branch, National Institute of Arthritis, Diabetes, Digestive and Kidney Diseases, Bethesda, Maryland

Cable Aspects of Dendrites • Relevance to Dendritic Synapses • Steady-State Voltage Decrement With Distance • Reversal Potential for Dendritic Synapse • Transient Solution for Voltage Step at Soma • Dendritic Transfer in AC and Laplace Transform Domains • Dendritic Modulus and Phase in AC Steady State • Relation to AC Impedances • Current Transfer From Dendrites to Clamped Soma • Transfer in Laplace Transform and Time Domains • Relation Between Brief Synaptic Current and Current Transient at Voltage Clamp • Discussion • Summary

The dendritic surface area of a neuron may be 10, 20, or even 100 times greater than the surface area of the neuron soma. This dendritic membrane provides a large distributed capacity that is electrically coupled to the soma by various cable distances along the branched core conductor. When a voltage clamp is applied across the soma membrane, the dendritic membrane is neither space clamped nor voltage clamped, except for the special case of very short dendritic trees.

For this chapter it is assumed that *1)* the microelectrode is inside a neuron soma, *2)* the soma membrane remains isopotential, and *3)* the control system applies ideal voltage clamping. Together these assumptions imply perfect space and voltage clamping of the soma membrane; thus attention can be focused on the cable properties of the dendritic trees.

Cable Aspects of Dendrites

The application of cable theory to dendritic trees has been presented, discussed, and reviewed in a book by Jack et al. (14) and in a chapter by

Rall (25); only brief reference is made to some of that literature here. First, it should be emphasized that the cable equations can be solved for dendritic trees with arbitrary branch lengths and diameters if the following simplifying assumptions are made: *1*) cylindrical branch elements, *2*) uniform passive membrane, *3*) uniform intracellular resistivity, *4*) extracellular isopotentiality, *5*) continuity of voltage, *6*) conservation of current at branch points, and *7*) a terminal boundary condition, usually taken to be a sealed end. The method of constructing the steady-state solution for such arbitrary branching was presented in 1959 (19) and has subsequently been applied to many neurons by many investigators. This approach to arbitrary branching was extended to AC steady states and transients by Barrett and Crill (2, 3), who also matched the changing diameter of noncylindrical branches with several short cylinders of different diameters. This approach was later used by S. Glasser and J. Miller (personal communication; see also ref. 30) and by D. Turner (personal communication; see also ref. 31). A different theoretical approach to arbitrary branching was provided by Butz and Cowan (6) and was later implemented by Horwitz (12, 13). Compartmental modeling provides an alternative computational approach; this can be implemented by means of computer programs designed to solve large systems of ordinary differential equations [(22, 23); see also ref. 18]. Although it was originally used to facilitate the specification of synaptic input at different dendritic locations (22, 23), this compartmental approach also permits different membrane properties and different inputs in selected regions of any specified compartmental-model representation of an arbitrarily branched neuron. Compartmental computations involving hundreds or thousands of compartments, chosen to match detailed dendritic anatomy, are currently being carried out by several research groups.

After establishing that arbitrary branching can be dealt with computationally, we chose to avoid these complications in this chapter. Very significant analytical advantage is gained by assuming a dendritic tree that can be represented by an equivalent cylinder of finite length with sealed ends. Although the tree in Figure 1 suggests symmetrical branching, it can have unequal branching provided that all terminal branches have sealed ends at the same electrotonic distance from the soma and that the branch diameters (raised to 3/2 power) yield the same sum at every electrotonic distance from the soma (14, 19, 21, 25). The electrotonic distance ($X = x/\lambda$) is a dimensionless variable that defines a mapping between the tree and the equivalent cylinder. In the tree, x represents the actual distance along the branches, whereas the length constant (λ) decreases with branching (proportional to square root of branch diameter for uniform materials). In the equivalent cylinder the value of λ remains constant and X ranges from 0 to $X = L = l/\lambda$, where

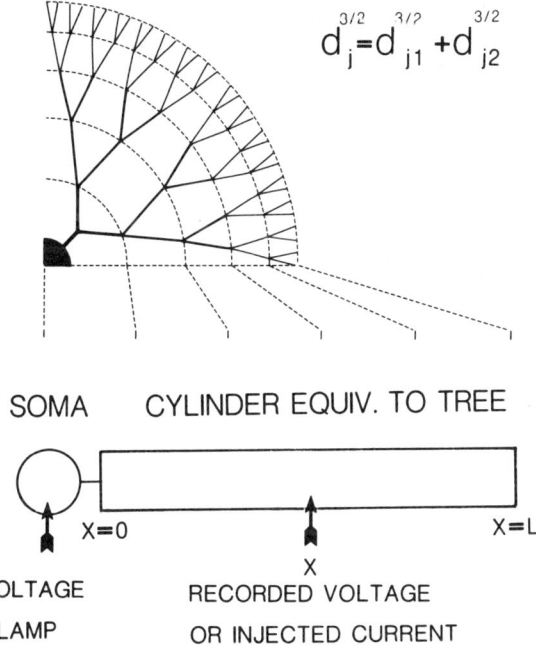

Fig. 1. *Top*: extensively branched tree attached to portion of neuron soma. *Bottom*: single cylinder (equivalent to tree) attached to lumped soma; cylinder extends from $X = 0$ to $X = L$. Voltage clamp is applied at soma, which is isopotential with cylinder at $X = 0$; dendritic location X is site of voltage recording or site of current injection.

l is the length of the equivalent cylinder and L is the dimensionless electrotonic length of both the tree and its equivalent cylinder.

Relevance to Dendritic Synapses

When voltage clamping of the soma membrane is used to study synaptic activity that is localized in the dendrites at some electrotonic distance away from the soma, it is critically important to distinguish the potential at the synaptic location from the clamping potential at the soma and to distinguish the current generated at the synaptic location from the current detected at the soma by the clamping circuit.

The current detected at the soma can be regarded as equal to the synaptic current only for the special case of a synapse located at the soma; otherwise the cable properties of the dendritic membrane cause both the time course and the amplitude of the actual synaptic current (at the synaptic input location) to differ from the transient current observed at the soma. The relation between these two transient currents can be expressed mathematically and computed numerically (see Eqs.

24–32 and Figs. 6 and 7). This relation has also been computed recently by means of compartmental simulations (D. Johnston and T. H. Brown, personal communication; see also refs. 5 and 16).

When a graduated series of clamping potentials is used to obtain reversal of the current transient detected at the soma for synaptic activation, the clamping potential that produces zero current is the true synaptic reversal potential only when the synapse is located at the soma; otherwise the zero current occurs when the clamping potential differs from rest by a larger amount than does the true reversal potential at the synaptic site. The relation between these two potentials is given by the DC steady-state solution (see Eq. 2). The usefulness of this simple result has been discussed at length by Jack et al. (14) and Carnevale and Johnston (8).

The importance of distinguishing the current (and potential) at the synaptic location from that at the soma was emphasized and illustrated some years ago (see Figs. 4 and 6 and Summary of ref. 23). Although specifically concerned with voltage transients rather than with voltage clamping, those compartmental computations were designed to distinguish explicitly between the synaptic conductance transient, the synaptic current transient, the loss current to neighboring compartments, the net depolarizing current at the synaptic location, the local voltage transient, the current transient reaching the soma compartment, and the resulting excitatory postsynaptic potential (EPSP) at the soma. They also dealt explicitly with the effects of steady-state hyperpolarization and depolarization applied at the soma and provided early insight into such problems (23).

Steady-State Voltage Decrement With Distance

When an equivalent cylinder is used to represent a dendritic tree, some people still jump to the erroneous conclusion that voltage decrement with distance is given simply by the exponential function $\exp(-x/\lambda)$. This function would be correct for a very long cylinder or for a short cylinder with a special boundary condition at the far end (25); however, it is not correct for a short cylinder with a sealed-end boundary condition.

For steady states the dimensionless form of the cable equation simplifies to the ordinary differential equation

$$d^2V/dX^2 - V = 0 \tag{1}$$

where $X = x/\lambda$ and V represents the departure of the intracellular potential from its resting value. The general solution of Equation 1 is a function of X containing two arbitrary constants; this can be expressed

$$V(X) = A\exp(-X) + B\exp(+X)$$

where the constants A and B are to be determined by two boundary conditions (see ref. 25 for remarks on relative merits of alternative expressions for this general solution). The sealed-end boundary condition $dV/dX = 0$ at $X = L$ implies that $B/A = \exp(-2L)$. Then the voltage-clamp boundary condition $V = V_o$ at $X = 0$ implies that $A = V_o/[1 + \exp(-2L)]$. The resulting solution of Equation 1 with these two boundary conditions can be expressed

$$V(X) = V_o \exp(-X) \frac{1 + \exp[-2(L - X)]}{1 + \exp(-2L)} \quad (2a)$$

or more compactly and in normalized form

$$\frac{V(X)}{V_o} = \frac{\cosh(L - X)}{\cosh L} \quad (2b)$$

This normalized decrement with X is illustrated in Figure 2 (solid curves) for three values of L. For comparison the dashed curve shows the simple exponential decrement corresponding to very large L.

At the distal end ($X = L$) this voltage ratio becomes simply

$$\frac{V(L)}{V_o} = \frac{1}{\cosh L} = \frac{2\exp(-L)}{1 + \exp(-2L)} \quad (3)$$

For the three L values in Figure 2, this gives 0.648, 0.425, and 0.266,

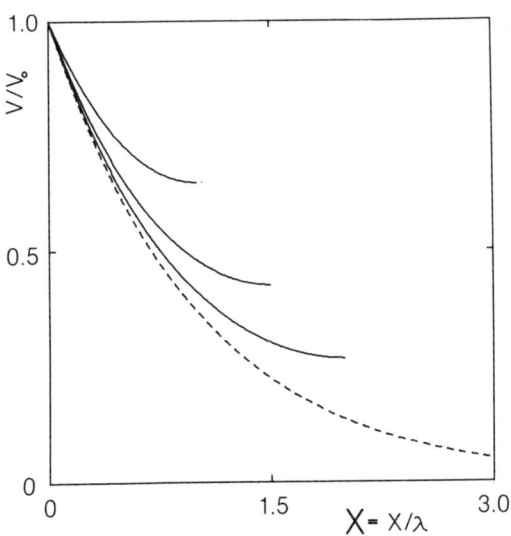

Fig. 2. Steady-state decrement of potential with distance ($X = x/\lambda$) in equivalent cylinder for 3 electrotonic lengths: $L = 1.0$, 1.5, and 2.0. *Dashed curve*, single exponential decrement corresponding to very large L values.

which are almost twice the corresponding values (0.368, 0.223, 0.135) of the dashed curve.

In contrast with the lack of space clamping seen in Figure 2, the case of a tree or equivalent cylinder of very short electrotonic length can be shown to result in approximate space clamping. By using the series expansion of $\cosh(L)$ or of $\exp(L)$, a useful approximation of Equation 3 (for small L) can be expressed

$$\frac{V(L)}{V_o} \cong \frac{1}{1 + 0.5L^2} \cong 1 - 0.5L^2 \qquad (4)$$

This means that for short L values of 0.1, 0.2, or 0.4, the steady-state departure from perfect space clamping would not exceed 0.5%, 2%, or 8%, respectively.

Reversal Potential For Dendritic Synapse

Let V_o represent the steady voltage-clamp value at the soma for which activation of a synapse located at electrotonic distance X produced a zero current transient at both the synaptic site X and the soma ($X = 0$). Note that the actual reversal potential at the synaptic site is not V_o but $V(X)$, which is smaller than (V_o) by a factor defined exactly by Equation 2. If X and L are known, then the value of $V(X)$ can be deduced from Equation 2 and V_o; if the reversal potential is known, then the value of X can be deduced from V_o and L. These matters are discussed at greater length by Jack et al. (14) and Carnevale and Johnston [(8); see also refs. 7 and 17].

Transient Solution For Voltage Step at Soma

When a voltage clamp is suddenly applied to the soma as a square voltage step, the early current flow charges mainly the capacitance of the nearest dendritic membrane; gradually both the voltage and the current reach out toward the distal dendritic membrane until the steady-state condition is attained. This spatiotemporal aspect of the early response to a voltage step at the soma is illustrated in Figure 3 for several different times but for only one length ($L = 1.5$). The steepest decrement shown is for $T = t/\tau = 0.1$; other curves are for $T = 0.2, 0.5, 1.0$, and the steady state [T is dimensionless; time (t) and passive membrane time constant (τ) are usually in ms]. These curves clearly show the extent to which voltage clamping at the soma fails to space clamp or voltage clamp the dendritic membrane.

The mathematical background for these theoretical results has previously been presented and discussed (24, 25). The mathematical solution [$V(X,T)$] of the cable equation in dimensionless form for a resting initial

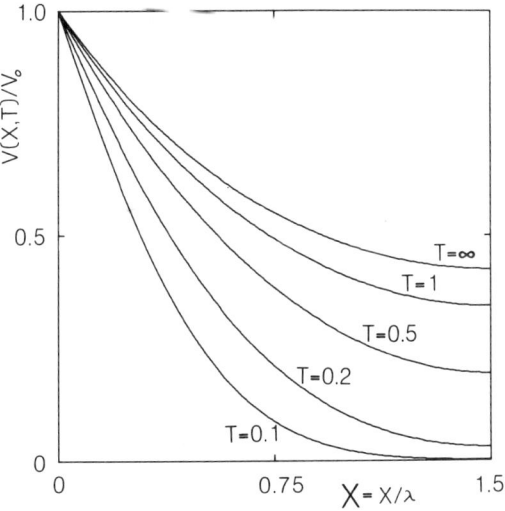

Fig. 3. Transient response $V(X,T)$ to voltage step applied at $X = 0$ and $T = 0$ for $L = 1.5$; computed from Eqs. 5 and 6 with n values from 1 to 10.

condition $[V(X,0) = 0]$ and for the two boundary conditions $[V(0,T) = V_o$ for the clamped end and $dV/dX = 0$ at $X = L$ for the sealed end] can be expressed

$$\frac{V(X,T)}{V_o} = \frac{\cosh(L - X)}{\cosh(L)} - \frac{2}{L} \sum_{n=1}^{\infty} \frac{\alpha_n \sin(\alpha_n X)}{(1 + \alpha_n^2)} \exp[-(1 + \alpha_n^2)T] \quad (5)$$

where

$$\alpha_n = (2n - 1)\pi/2L \quad (6)$$

and n has integer values from 1 to ∞. The first term on the right side of Equation 5 is simply the steady state of Equation 2b, whereas the infinite series represents a sum of exponential decays. If each exponential is expressed as $\exp(-t/\tau_n)$, these equations imply decay time constants whose relation to the passive membrane time constant (τ_m) can be expressed

$$\tau_m/\tau_n = 1 + \alpha_n^2 = 1 + [(2n - 1)\pi/2L]^2 \quad (7)$$

For $L = 1.5$, Equation 7 has values of ~2.1, 10.9, 28.4, and 54.7 corresponding to n values of 1, 2, 3, and 4, respectively. These time-constant ratios are significantly different from those obtained in the absence of the voltage clamp. For comparison the cylinder with both ends sealed has $\alpha_n = n\pi/L$, including $n = 0$; the values of α_n become more complicated when the soma is explicitly included at $X = 0$ without voltage clamping (24).

Because the current flowing from the clamp into this dendritic tree at $X = 0$ is proportional to the negative gradient of V at $X = 0$, it is useful to express this in normalized form

$$\left.\frac{(\partial V/\partial X)}{V_o}\right|_{X=0} = \tanh(L) + \frac{2}{L}\sum_{n=1}^{\infty}\frac{\alpha_n^2}{1+\alpha_n^2}\cos(\alpha_n X)\exp[-(1+\alpha_n^2)T] \quad (8)$$

where $\tanh(L)$ corresponds to the current needed to sustain the eventual steady state, whereas the infinite series corresponds to the additional current needed to charge the dendritic membrane during the early (dendritic transient) aspect of voltage clamping at $X = 0$.

On the other hand, differentiation of Equation 5 with respect to T and normalization with respect to V_o gives

$$\frac{(\partial V/\partial T)}{V_o} = \frac{2}{L}\sum_{n=1}^{\infty}\alpha_n \sin(\alpha_n X)\exp[-(1+\alpha_n^2)T] \quad (9)$$

This is proportional to the response in the limiting case of a Dirac δ-function voltage clamp at $X = 0$; this can be thought of as a voltage step from 0 to V_o at $X = 0$ and $T = 0$, followed by a voltage step back down from V_o to 0 at $X = 0$ and $T = \Delta T$ in the limit as $\Delta T \to 0$. This function provides a useful representation in the time domain, related to a Laplace transform transfer function (see Eqs. 27–32).

Dendritic Transfer in AC and Laplace Transform Domains

Both for AC steady states and for the Laplace transform domain, the cable equation for a uniform membrane cylinder reduces to the ordinary differential equation

$$d^2\hat{V}/dX^2 - q^2\hat{V} = 0 \quad (10)$$

where $\hat{V}$ and q are both complex quantities (with real and imaginary parts, implying a modulus and a phase angle). In the AC (sinusoidal steady state) domain, $q^2 = 1 + j\omega\tau$, where $j = \sqrt{-1}$ and ω is the angular frequency $2\pi f$. In the Laplace transform domain, $q^2 = 1 + \tau s$, where the complex variable s (sometimes termed *complex frequency*) is used to define the relation of $\hat{V}(X,s)$ in the Laplace transform domain to $V(X,t)$ in the time domain (see Eq. 26; Mathematical appendix of ref. 14, or see ref. 9, where p is used instead of s).

Equation 10 is similar to Equation 1 and has a general solution that is similar except that A and B are now complex arbitrary constants, the argument X is replaced by qX, and $V(X)$ is replaced by $\hat{V}(X,q)$. When the sealed-end boundary condition is generalized to $d\hat{V}/dX = 0$ at $X = L$

and the clamping condition is generalized to an imposition of $\hat{V}(0,q)$ at $X = 0$, then the solution (corresponding to Eq. 2b) becomes generalized to

$$\frac{\hat{V}(X,q)}{\hat{V}(0,q)} = \frac{\cosh[q(L - X)]}{\cosh(qL)} = \hat{F}(X,q) \qquad (11)$$

The interpretation of this ratio, which is sometimes called a transfer function, is complicated by the presence of q in the arguments of the hyperbolic cosines; this complication is treated differently in the two domains.

Dendritic Modulus and Phase in AC Steady State

For AC steady states, q^2 has the physical meaning of membrane admittance divided by membrane conductance

$$q^2 = \frac{G_m + j\omega C_m}{G_m} = 1 + j\omega\tau \qquad (12)$$

where the membrane conductance (G_m) and the membrane capacitance (C_m) are both per unit area, and their ratio (C_m/G_m) equals the passive membrane time constant (τ or τ_m). Note that $\omega\tau = 2\pi f\tau$ is a dimensionless parameter proportional to frequency; it equals π exactly for f = 100 Hz, with τ = 5 ms. Also, for $\omega\tau = \pi$ the modulus of q is close to 1.82 with a phase angle of 0.63 radians, implying a real part of 1.47 and an imaginary part of 1.07 (for verification see Eqs. A1–A5 in **Appendix A**, p. 211).

For each value of $\omega\tau$, the ratio defined by Equation 11 can be expressed for an AC steady state as

$$\frac{\hat{V}(X,\omega\tau)}{\hat{V}(0,\omega\tau)} = \frac{\cosh(aY + jbY)}{\cosh(aL + jbL)} = M \exp(j\phi) \qquad (13)$$

where $Y = L - X$, a and b represent the real and imaginary parts of q, respectively, and M and ϕ represent the modulus and phase angle of this complex ratio, respectively. From **Appendix A** it follows that

$$M = \left[\frac{\cosh(2aY) + \cos(2bY)}{\cosh(2aL) + \cos(2bL)}\right]^{1/2} \qquad (14)$$

and that

$$\phi = \arctan[\tanh(aY)\tan(bY)] - \arctan[\tanh(aL)\tan(bL)] \qquad (15)$$

These quantities have been computed and plotted in Figures 4 and 5 as functions of X for L = 1.5 and for $\omega\tau$ values chosen as multiples of π (corresponding to same multiples of 100 Hz if τ = 5 ms).

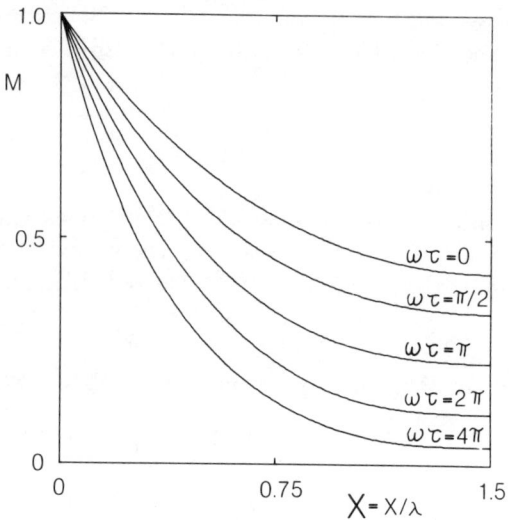

Fig. 4. AC steady-state modulus versus X for $L = 1.5$; computed from Eq. 14 for frequencies ($\omega\tau$) shown, corresponding to 400, 200, 100, 50, and 0 Hz (from *bottom curve* to *top curve*) if $\tau = 5$ ms.

Figure 4 shows clearly how the modulus of AC voltage decreases with increasing distance from the point of AC voltage clamping (at $X = 0$) to the dendritic terminals (at $X = L = 1.5$). The upper curve (for $\omega\tau = 0$) is simply the DC steady-state decrement shown in Figures 2 and 3. The other curves (for values of $\pi/2$, π, 2π, and 4π) show steeper decrement at left and level off to lower values at right.

Useful approximations for the slope at $X = 0$ (at different frequencies) can be obtained by noting that $a < 1$ for all frequencies, implying that $2aL < 3$ (for $L = 1.5$) and that both $\cosh(2aL)$ and $\sinh(2aL)$ are >10, whereas both sine and cosine never exceed unity. Thus the exact and approximate values for the slope at $X = 0$ can be expressed

$$\frac{dM}{dX}\bigg|_{X=0} = -a\,\frac{\sinh(2aL) - (b/a)\sin(2bL)}{\cosh(2aL) + \cos(2bL)} \sim -a \quad (16)$$

For the $\omega\tau$ values used in Figure 4, the a values are ~1, 1.2, 1.47, 1.92, and 2.61, respectively (see Eq. A4). Thus, for example, a quick estimate of the value of M can be obtained at $X = 0.1$, as $(1 - 0.1a) = 0.88$ for $\omega\tau = \pi$. This also suggests the approximation (for small X)

$$M \cong \exp(-aX) \quad (17)$$

which becomes exact when L is very large (as could be derived directly for cylinder of semi-infinite length).

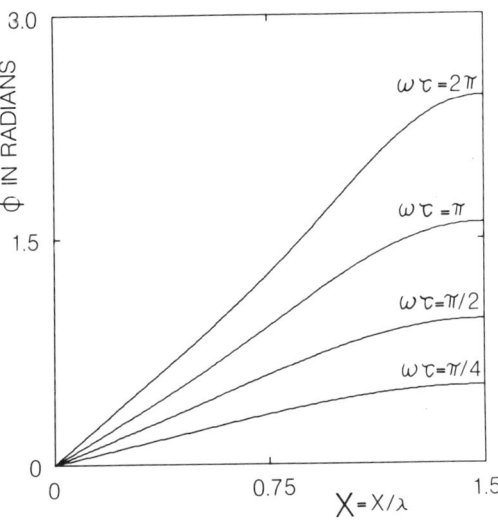

Fig. 5. AC steady-state phase lag versus X for $L = 1.5$; computed from Eq. 15 for frequencies ($\omega\tau$) shown, corresponding to 25, 50, 100, and 200 Hz (from *bottom curve* to *top curve*) if $\tau = 5$ ms.

The exact and approximate values of M at $X = L$ can be expressed

$$M_{X=L} = \left(\frac{2}{\cosh(2aL) + \cos(2bL)}\right)^{1/2} \cong 2\exp(-aL) \qquad (18)$$

For $L = 1.5$ and the $\omega\tau$ values in Figure 4, the approximate values are 0.44, 0.33, 0.22, 0.11, and 0.04 for M at $X = L$. This provides a quick and simple estimate of the failure of space clamping at the dendritic terminals (as function of frequency and L).

Figure 5 shows that the phase lag $(-\phi)$ of Equation 15 increases monotonically with X; it also increases with the AC frequency. The values used for the curves in Figure 5 were $\pi/4$, $\pi/2$, π, and 2π.

To gain insight into these results, we seek approximate expressions for slopes and values near $X = 0$ and $X = L$. At $X = L$, $Y = L - X = 0$, and the contribution of the Y-dependent term in Equation 15 is zero; thus the phase lag at $X = L$ can be expressed exactly and approximately as

$$-\phi_L = \arctan[\tanh(aL)\tan(bL)] \cong bL \qquad (19)$$

where b is defined by Equation A5. The approximate value of Equation 19 is exact for the particular bL values (0 and $n\pi$) for which $\tan(bL) = 0$, and for bL values that are odd multiples of $\pi/2$ for which $\tan(bL) = \pm\infty$; however, for bL values intermediate to these, the approximate value differs when tanh(aL) differs significantly from unity (<2% for $aL \geqq$ 2.3).

In Figure 5 the phase lags shown at $X = L$ are 0.52, 0.96, 1.61, and 2.46 radians, which agree quite well with the approximate bL values of 0.55, 0.98, 1.61, and 2.46 radians. For $\omega\tau = \pi$ (i.e., 100 Hz for $\tau = 5$ ms) the phase lag is close to $\pi/2$; this means that the AC voltage at the dendritic terminals is close to 90° out of phase with the AC voltage at the soma for this case. For small values of X the variable ($Y = L - X$) has values close to L; thus the Y-dependent term in Equation 15 can be approximated similarly (as Eq. 19). This yields the expression (for small X)

$$-\phi \approx b(L - Y) \approx bX \tag{20}$$

which becomes exact for all X when L is very large (as could be derived directly for cylinder of semi-infinite length); this pairs with Equation 17 for M. A quick estimate of the phase lag can be obtained at $X = 0.1$ as $0.1b$ or 0.011 radians for $\omega\tau = \pi$.

As $X \to L$, the variable ($Y = L - X$) becomes small and we make the approximation $\arctan[\tanh(aY)\tan(bY)] \cong abY^2 \cong (\omega\tau/2)(L - X)^2$. Combination of this result with Equation 19 yields the useful approximate expression (for values of X close to L)

$$-\phi \cong bL - (\omega\tau/2)(L - X)^2 \tag{21}$$

This accounts for the curvature seen in Figure 5 as $X \to L$. The slope near $X = L$ can be expressed

$$d(-\phi)/dX \cong \omega\tau(L - X) \tag{22}$$

which becomes zero when $X = L$ (as it should, because sealed-end boundary condition at dendritic terminals implies zero derivative for both real part and imaginary part of $\hat{V}$).

Relation to AC Impedances

When AC current is applied at $X = 0$ for a time sufficient to produce a sinusoidal steady state of the entire system, the input impedance can be expressed $\hat{Z}_{o,o} = \hat{V}(0,q)/\hat{I}_A(0,q)$ and the transfer impedance to the dendrites at distance X can be expressed $\hat{Z}_{o,X} = \hat{V}(X,q)/\hat{I}_A(0,q)$, where $\hat{V}(0,q)$ and $\hat{V}(X,q)$ represent the AC voltage at $X = 0$ and at distance X, respectively, whereas $\hat{I}_A(0,q)$ represents the AC current applied at $X = 0$. Because both impedance expressions have the same denominator, the ratio of these complex impedances

$$\hat{Z}_{o,X}/\hat{Z}_{o,o} = \hat{V}(X,q)/\hat{V}(0,q) = \hat{F}(X,q) \tag{23}$$

exactly equals the ratio defined by Equations 11–13. Consequently all of the results from Equations 14–22 and Figures 4 and 5 apply equally well to this impedance ratio. The individual impedances are more complicated because the applied current drives both the soma and the dendrites (14, 15, 20, 28, 29).

Current Transfer From Dendrites to Clamped Soma

When AC current is injected in the dendrites at location X during a previously established steady DC voltage clamp at $X = 0$, the ratio of $\hat{I}_c(0,q)$ (AC current detected by clamp at $X = 0$) to $\hat{I}_{in}(X,q)$ (injected AC current) equals the previous ratio (Eq. 11), i.e.

$$\frac{\hat{I}_c(0,q)}{\hat{I}_{in}(X,q)} = \frac{\cosh[q(L-X)]}{\cosh(qL)} = \hat{F}(X,q) \qquad (24)$$

as demonstrated in **Appendix B** (p. 212). This is a generalization of a similar result obtained by Carnevale and Johnston (8) for DC current injection in terms of a resistance ratio symbolized as their coupling coefficient, k_{12}. Subsequently Johnston and Brown (personal communication and ref. 16) demonstrated such a result for the modulus of AC current in a compartmental-model simulation.

The point to be emphasized here is that for AC current injection in the dendrites at any X and any frequency the attenuation of amplitude and the shift in phase of the current detected at the soma clamp are defined by Equations 11–22. This provides explicit analytical expressions for correcting the distortion expected in the AC current recorded at the soma clamp.

This result holds, no matter how the injected current is divided between several branches of the dendritic tree, provided that each component is injected at the same value of X (and each has identical frequency and phase); this generalization is a consequence of the superposition properties of such trees, as was explained in collaboration with Rinzel (27, 29). Furthermore different inputs of different frequencies could be injected at different electrotonic locations, and the resultant current at the soma clamp would be the superposition of the separate effects of the individual inputs.

Transfer in Laplace Transform and Time Domains

Looking back to Equations 10–12, a different expression for q^2 is now used, i.e.

$$q^2 = 1 + \tau s \qquad (25)$$

where the variable s can assume complex values. Then the functions $\hat{V}(X,q)$, $\hat{I}(X,q)$, and $\hat{F}(X,q)$ of Equations 11 and 24 can be reexpressed as functions of s (in Laplace transform domain); these functions $\hat{V}(X,s)$,

$\hat{I}(X,s)$, and $\hat{F}(X,s)$ are related to the corresponding functions $V(X,t)$, $I(X,t)$, and $F(X,t)$ in the time domain by the Laplace transformation. This transformation can be defined and expressed

$$\hat{F}(X,s) = \mathscr{L}[F(X,t)] = \int_0^\infty \exp(-st)F(X,t)\,dt \qquad (26)$$

whereas the inverse transformation is defined and expressed

$$F(X,t) = \mathscr{L}^{-1}[\hat{F}(X,s)] = \mathscr{L}^{-1}\left\{\frac{\cosh[q(L-X)]}{\cosh(qL)}\right\} \qquad (27)$$

Thus Equation 27 serves to define $F(X,t)$ as the impulse response function (in time domain) corresponding to the particular transfer function $\hat{F}(X,s)$ or $\hat{F}(X,q)$ specified by Equation 11 (in Laplace transform domain). For a useful discussion of Laplace transforms and their inverses, see the mathematical appendix of reference 14, which also explains and discusses the convolution integral (with respect to variable u, which runs from 0 to t); thus

$$V(X,t) = V(0,t)*F(X,t) = \int_0^t V(0,t-u)F(X,u)\,du \qquad (28a)$$

where the expression with the asterisk represents a compact symbolic notation for the convolution integral shown at the right. This convolution in the time domain corresponds to a simple product in the Laplace transform domain

$$\hat{V}(X,q) = \hat{V}(0,q)\hat{F}(X,q) \qquad (28b)$$

Note that $V(X,t)$ is the inverse Laplace transform of $\hat{V}(X,q)$ and that Equation 28b represents a rearrangement of Equation 11.

In other words, when a voltage clamp at $X = 0$ is used to impose a transient voltage $V(0,t)$, the resulting voltage transient $V(X,t)$ at distance X can be calculated as the convolution (Eq. 28a) of $V(0,t)$ with the impulse response function $F(X,t)$ or by inverse Laplace transformation of the product in Equation 28b.

Also, when a brief current transient $I_{in}(X,t)$ is injected in the dendrites at X during a previously established steady DC voltage clamp at $X = 0$, the transient current $I_c(0,t)$ (detectable by clamp) can be predicted theoretically (see **Appendix B**, p. 212) as the convolution

$$I_c(0,t) = I_{in}(X,t)*F(X,t) \qquad (29)$$

Expressions for $F(X,t)$ can be obtained in several ways. Because the solution for a voltage step applied at $X = 0$ is available (Eq. 5), the simplest way is differentiation of that solution with respect to t; thus

Space-Clamp Problems in Branched Neurons

$$F(X,t) = \frac{2}{\tau L} \sum_{n=1}^{\infty} \alpha_n \sin(\alpha_n X) \exp[-(1 + \alpha_n^2)t/\tau] \tag{30}$$

where α_n is defined by Equation 6. To verify this note that, for a voltage step at $X = 0$ [i.e., $V(0,t) = V_o H(t)$ in Eq. 28a, where $H(t)$ is the Heaviside step function], $V(X,t)$ equals the integral of $F(X,t)$ with respect to t; differentiating with respect to t gives $dV/dt = F(X,t)$.

A different expression for $F(X,t)$ can be obtained by reference to the G functions described by Horwitz (see Eqs. 29–34 of ref. 12 with $m = 0$ and $n = 1$; also note that symbols γ, L, and D of ref. 12 correspond to q/λ, λL, and λX, respectively, in this chapter). When $T = t/\tau$ is used to simplify the expressions, the result is

$$F(X,T) = \sum_{k=0}^{\infty} (-1)^k \{g(2kL + X, T) + g[2(k+1)L - X, T]\} \tag{31}$$

where

$$g(Z,T) = \frac{ZT^{-3/2}}{2\tau\sqrt{\pi}} \exp(-T - Z^2/4T) \tag{32}$$

noting that $g(Z,T)$ has the Laplace transform

$$\mathscr{L}[g(Z,T)] = \exp(-Z\sqrt{\tau s + 1}) \tag{33}$$

Closely related inverse Laplace transforms have been presented and discussed in neuronal context by Jack, Redman, and co-workers (14, 15, 28) and by Rinzel and Rall (29).

Computations with both expressions (Eqs. 30 and 31) were carried out to provide Figure 6 and to verify that these two rather different expressions define the same impulse response function. It should be emphasized that $F(X,t)$ is an impulse response function whose meaning differs from the more familiar examples [which define the voltage transient at 1 location in response to instantaneous charge injected at another point; (e.g., see ref. 29)]; here it defines a current transient at one location (point of voltage clamp, $X = 0$) in response to an instantaneous charge injected at another point (X), or it defines a voltage transient at X in response to a temporal Dirac δ-function of voltage at $X = 0$ (note explanation after Eq. 9). Thus Equation 28 defines $V(X,t)$ when an arbitrary voltage transient $V(0,t)$ is imposed at $X = 0$; in the other case, Equation 29 defines the transient current detected by a steady DC voltage clamp at $X = 0$ when an arbitrary current transient $I_{in}(X,t)$ is injected at X.

Relation Between Brief Synaptic Current and Current Transient at Voltage Clamp

These theoretical results are highly relevant to the interpretation of experimental records showing the current transient at the voltage clamp

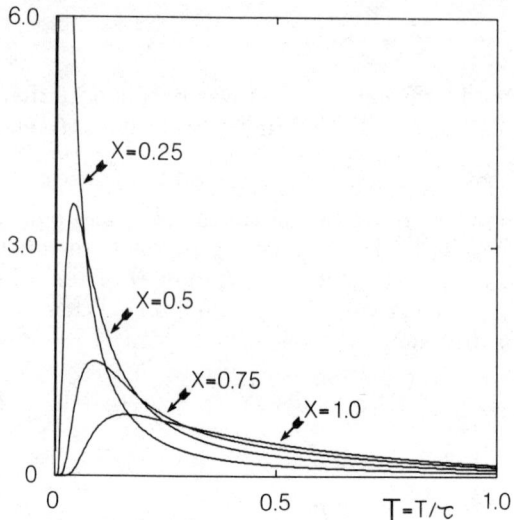

Fig. 6. Plot of τ times transfer function $F(X,T)$ (defined by Eqs. 30–32) versus T for X values of 0.25, 0.50, 0.75, and 1.0 with $L = 1.5$; limiting case (for $X = 0$) is Dirac δ-function. This gives $V(X,T)$ in response to voltage clamp at $X = 0$; it also gives current detected by voltage clamp at $X = 0$ for δ current injection at X.

for brief discrete synaptic input (5, 10, 16). Given the linearity and other simplifying assumptions stated at the beginning of this chapter, setting the values of X and L determines $F(X,t)$. The convolution (Eq. 29) defines the forward process of getting from the input-current time course $I_{in}(X,t)$ to the current transient $I_c(0,t)$ at the somatic voltage clamp. Given a restriction to brief input at a single location, this relation determines two inverse problems: *1*) estimate the input time course $I_{in}(X,t)$ when given X, L, and $I_c(0,t)$ and *2*) estimate the value of X when given L, $I_c(0,t)$, and $I_{in}(X,t)$. The first problem can probably be solved by a numerical deconvolution (this is currently being explored); in any case, such a numerical result can always be tested by forward convolution to determine whether it is valid. However, because of interest in the second problem, it is simpler and probably more practical to facilitate both problems by computing a set of shape indices [as was done for EPSP shapes; (see refs. 14, 15, 23, 25)] and plotting their dependence on X, L, and the input shape parameter α. The input time course is assumed to be determined by a single parameter (α) in the expression

$$I_{in}(T) = Q\alpha^2 T \exp(-\alpha T) \tag{34}$$

where $T = t/\tau$ and Q is the input charge corresponding to the area (under curve) found by integrating $I_{in}(T)$ from $T = 0$ to $T = \infty$; the peak current occurs when $T = T_p = 1/\alpha$.

In Figure 7 the fastest curve corresponds to $\alpha = 50$. It represents both $I_{in}(X,T)$ and $I_c(0,T)$ when $X = 0$. The other curves illustrate the result of numerical convolution of this same input time course with $F(X,T)$ of Equation 31 for the particular X values 0.1, 0.25, 0.5, 0.75, and 1.0 and with $L = 1.5$ in each case. These curves represent the predicted amplitude and time course of $I_c(0,T)$ as the input current is shifted to successive dendritic locations.

From these results it is clear that, even when the input location is as close to the soma as $X = X/\lambda = 0.1$ (for this case of $\alpha = 50$ and $L = 1.5$), the current transient to be detected by the somatic voltage clamp is significantly changed from the time course of the input current; i.e., the peak amplitude is reduced by ~33% and the peak time is increased by ~60%; also the half-width is increased by ~40% and the slope (halfway down) corresponds to an apparent current decay time constant that is increased by ~30%.

The six relative peak amplitudes in Figure 7 are 1.0, 0.657, 0.358, 0.155, 0.078, and 0.044; the six peak times are 0.02, 0.032, 0.05, 0.09, 0.14, and 0.21, as read from the numerical listing of the computed results. Additional computations with different values for α and L are being done. The analysis and summary of those results will be presented when completed.

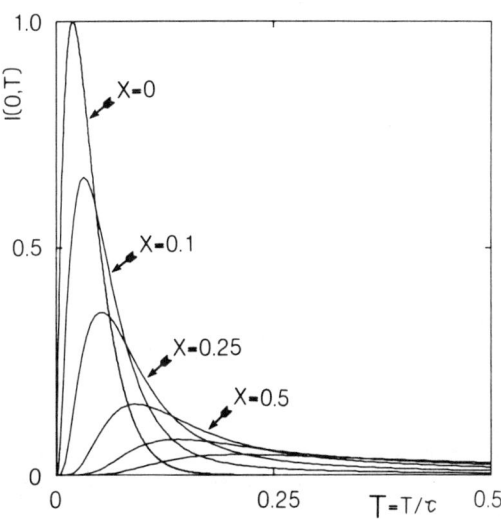

Fig. 7. Predicted current detected by voltage clamp at $X = 0$ for brief transient current injection at X; input current defined by Eq. 34 for $\alpha = 50$. Computed results obtained by convolution (Eq. 29) of current with $F(X,T)$ from Eq. 31. Curve with earliest peak corresponds to $X = 0$ and was actually computed for $X = 0.01$; 2 unlabeled curves (with latest peaks) correspond to $X = 0.75$ and 1.0; $L = 1.5$ in all cases.

Discussion

The theoretical results presented in this chapter depend critically on the assumption of passive (linear) membrane properties. Small nonlinearities, with complete absence of excitability (e.g., rectification), could have small effects that might be studied by means of perturbation-theory corrections to the present results. However, larger nonlinearities, especially when membrane excitability properties are present in some regions of the extended membrane surface, could not be treated as minor perturbations of linear results. Once specified, such complications can be studied numerically with a compartmental model that includes these specifications.

The existence of superposition properties (for current injection at different locations) and reciprocity properties (for transfer impedances between 2 points or for voltage transfer and current transfer described in this chapter) is a well-known physical and mathematical consequence of assuming a linear system. A dendritic tree need not be equivalent to a cylinder to make use of such properties. These properties provide the basis for the paper by Carnevale and Johnston (8); for DC coupling coefficients these authors pointed out that one measurement permits the prediction of related ratios without assuming any particular branching geometry (provided system is linear). The generalization of these ratios to AC steady states and to transfer functions in the Laplace transform domain is equally valid. An important implication of such general properties is that, as improved experimental techniques make it possible to measure AC steady states and transients at both somatic and dendritic locations, these reciprocity predictions provide criteria for testing the linearity of the system.

The advantages of explicit mathematical expressions for the dependence of these functions on X and L, with q and $\omega\tau$ or with τs and t, can be obtained by treating trees as cylinders. This chapter provides several such explicit results, including graphical illustrations.

These results clearly show that the dendritic membrane is neither space clamped nor voltage clamped, except for trees of very short electrotonic length. A short L value of 0.2 would ensure that departure from perfect space and voltage clamping is <2% for the DC steady state (Eq. 4) and for AC steady states at low frequencies ($\omega\tau < \pi$ in Eq. A11). However, for a higher frequency ($\omega\tau = 10\pi$, corresponding to 1 kHz with $\tau = 5$ ms and implying $a = 4.027$ and $b = 3.9$), $L = 0.2$ would imply that $\cosh(2aL) = 2.6$, $\cos(2bL) = 0.01$, and from Eq. 18 the departure would exceed 12% (for $\omega\tau = 50\pi$, departure would exceed 65%).

For information on early experiments and a discussion of voltage clamping the soma of a dendritic neuron see references 1, 4, 11, and 17.

More recently Johnston and Brown (personal communication and ref. 16) carried out both analog and digital simulations to aid in their interpretation of voltage-clamp measurements in hippocampal neurons. They assumed passive membrane properties and used compartmental models corresponding to $L = 0.9$ for dendrites represented as an equivalent cylinder. They reported results of various useful control experiments with these models and included curves showing voltage attenuation with distance for several AC frequencies; these curves are quite similar to the theoretical curves shown in Figure 4. They also showed that very similar curves can be obtained by comparing clamp current with AC current injected at dendritic locations (see Eq. 24). They concluded that the cable properties of the dendrites (and not voltage-clamp circuit) provide the limiting factor for voltage clamping of events that are not located in the cell soma. For their nearest dendritic compartment, corresponding to $X = 0.06$, they reported a bandwidth of 300 Hz (which they found barely adequate for reported synaptic current time course). For $X > 0.1$ the bandwidth was much smaller and serious distortion was expected; reliable synaptic current measurements were not expected. They also included simulations treating synaptic input as a conductance transient, as was done earlier by Rall (21–23) and by Barrett and Crill (3). They did not present the possibility of deducing the synaptic current transient from the convolution relation presented and discussed with Equations 29–34.

Finkel and Redman (personal communication and ref. 10) selected group Ia single-axon inputs (to cat spinal motoneurons), which produce EPSP shape indices suggesting a somatic location of the synapses. For a somatic location the current detected by the voltage clamp at the soma is correctly reported as synaptic current, except for clamping-error corrections, which they discuss. However, for input locations that are not quite somatic, corrections based on Equations 29–34 should be considered. For example, if the input location is on the dendritic trunks at $X = 0.1$ (and $L = 1.5$), Figure 7 shows that the true synaptic current peaks earlier than that recorded by the voltage clamp at the soma. For this particular example ($\alpha = 50$) the true peak time is about two-thirds of that observed and the true peak amplitude is ~50% greater than that observed; also, the true values for the half-width and the apparent current-decay time constant are both smaller (between 20% and 30%) than those observed.

Such measurements and calculations with synaptic currents have two important advantages over the earlier focus on EPSP time course and amplitude. *1)* With voltage clamping at the soma, the value of ρ (ratio of dendritic to somatic steady-state conductance) does not need to be known (cf. ref. 24); this contrasts with EPSP amplitude and time course for which the value of ρ is important even though it often cannot be accurately determined. The voltage clamp avoids all this uncertainty. 2) It is not necessary to know whether this input current is on a branch of large diameter or small diameter; the current recorded at the soma will be the same even when this input current is divided between several branches, provided that the several input currents are simultaneous, have the same α, and are located at the same X value (in the same tree or even in different trees with same L value). However, to calculate the voltage transient at each synaptic site, one needs to know how the input current is divided between branches of different diameter. As discussed by Redman (28) and by Rinzel and Rall (29), the response function at the input site depends on branch diameter and determines the local voltage generated by a given current injection. These local voltages are important when considering the nonlinearities that result from treating synaptic input as a conductance transient (3, 26, 28, 29).

Note that the presence of a voltage clamp at the soma has the effect of decoupling (from each other) the several dendritic trees belonging to the same neuron; activity in one tree has no effect on the other trees and each tree is coupled only to the voltage clamp. The several trees can have different L values; the solution within each tree depends only on the voltage-clamped boundary condition at $X = 0$, the sealed end at $X = L$, and the input received by that tree. Current generated in any tree can spread to the soma but not to the other trees. The current recorded by the clamp at the soma is a linear combination of the several currents supplied by the several trees (which may have different inputs and L values). Simple interpretations depend on simple input to only one tree.

Summary

1. Assumming perfect space and voltage clamping of the soma, the focus is on a cable-theory treatment of how seriously space and voltage clamping fail in the dendrites. The linear properties of passive membrane are assumed (see **Discussion**, p. 208).

2. The DC steady state and low-frequency AC steady states are space clamped within a 2% error only for very short trees with electrotonic lengths <0.2; much larger errors occur at higher frequencies and for larger L values.

3. The DC steady-state solution provides the basis for estimating the dendritic synaptic equilibrium potential from the reversal potential observed experimentally at the soma.

4. The AC steady-state solution provides formal expressions and convenient approximate expressions for the decrement of voltage amplitude with distance at different frequencies (see Fig. 4) and for the increasing phase lag with distance at different frequencies (see Fig. 5).

5. The transient response $V(X,T)$ of the dendritic membrane for a

voltage step applied at the soma is provided mathematically in Equation 5. The distal portion of the dendritic tree charges less rapidly than the proximal portion (see Fig. 3).

6. When either an AC or a brief transient current is injected at a dendritic location while a steady-state DC voltage clamp is maintained at the soma, the AC or transient current detected by the clamping circuit at the soma can be predicted mathematically and computed numerically (see Figs. 6 and 7). This is expressed by Equations 24 and 29 and verified in **Appendix B** (p. 212).

7. The computations for Figure 6 verified that Equations 30–32 provide two different expressions for the same transfer function. This transfer function has two physical meanings: *1)* through Equation 27 it yields $V(X,t)$ at any dendritic location when an arbitrary voltage time course $V(0,t)$ is imposed by voltage clamp at $X = 0$ and *2)* through Equation 29 it yields the current $I_c(0,t)$ detected by a previously established DC voltage clamp (at $X = 0$) when an arbitrary current transient $I_{in}(X,t)$ is injected at a dendritic location X.

8. Computations like those used to obtain Figure 7 provide a theoretical basis for deducing the time course of synaptic current (at particular electrotonic distance at which it is generated) from the current detected by voltage clamp at the soma. It may prove feasible to do this by a numerical deconvolution; in any case it can be done with shape indices and testing with a forward convolution. The true input current has a larger peak, shorter time to peak and half-width, and faster apparent decay time constant than that detected by the clamp at the soma.

9. The several dendritic trees originating from a common soma become electrically decoupled from each other when the soma is voltage clamped (see **Discussion**, p. 208).

Appendix A

For AC steady states the cable equation and its solution in the present context have been presented in Equations 10–15. The definitions, which both clarify and facilitate manipulations involving the real and imaginary parts of q, are summarized

$$q = a + jb = (1 + j\omega\tau)^{1/2} = r^{1/2} \exp(j\theta/2) \tag{A1}$$

$$r = (1 + \omega^2\tau^2)^{1/2} = \text{modulus of } q^2 \tag{A2}$$

$$\theta = \arctan(\omega\tau) = \text{phase of } q^2 \tag{A3}$$

$$a = r^{1/2} \cos(\theta/2) = [(r + 1)/2]^{1/2} \tag{A4}$$

$$b = r^{1/2} \sin(\theta/2) = [(r - 1)/2)]^{1/2} \tag{A5}$$

These definitions correspond to those in previous appendices by Rall (20) and by Rall and Rinzel (27). The hyperbolic cosine of complex argument can be expressed

$$\cosh(x + jy) = \cosh(x) \cos(y) + j \sinh(x) \sin(y)$$

From this it follows that the phase angle (ϕ_L) and the modulus (M_L) of $\cosh(qL)$ can be expressed

$$\phi_L = \arctan[\tanh(aL)\tan(bL)] \tag{A6}$$

$$M_L = [\cosh^2(aL)\cos^2(bL) + \sinh^2(aL)\sin^2(bL)]^{1/2} \tag{A7}$$

Also, because of standard identities involving the squares of sine, cosine, sinh, and cosh, there are several useful alternative expressions for M_L such as

$$M_L = [\cosh^2(aL) - \sin^2(bL)]^{1/2} \tag{A8}$$

$$= [(\tfrac{1}{2})\cosh(2aL) + (\tfrac{1}{2})\cos(2bL)]^{1/2} \tag{A9}$$

When L is replaced by $Y = L - X$, similar expressions hold for ϕ_Y and M_Y. Then from Equation 13

$$M \exp(j\phi) = \frac{\cosh(qY)}{\cosh(qL)} = \frac{M_Y}{M_L} \exp[j(\phi_Y - \phi_L)] \tag{A10}$$

and it can be seen that $M = M_Y/M_L$, explaining Equation 14, and that $\phi = \phi_Y - \phi_L$, explaining Equation 15.

For very short dendritic trees, L is small; for low and moderate frequencies, aL and bL are small and the series expansions for $\cosh(aL)$ and $\sin(bL)$ can be used to obtain a useful approximation from Equation A8. Thus

$$(M_L)^2 \cong (1 + 0.5a^2L^2)^2 - b^2L^2$$

$$\cong 1 + L^2(1 + 0.25a^4L^2)$$

implying

$$M_L \cong 1 + 0.5L^2[1 + (a^2L/2)^2] \tag{A11}$$

for small L and aL. For example, $L = 0.2$ and $\omega\tau = \pi$ imply a value of 1.021 for M_L or a value of -0.98 for the modulus of Equations 13 and 14 at $X = L$. This means that (for this frequency) very short dendritic trees corresponding to $L = 0.2$ would be space clamped and voltage clamped within 2.1%.

Appendix B

This section derives the solution for an AC steady state when AC current is injected at $X = A$, with voltage clamping at $X = 0$ and a sealed end at $X = L$. This AC steady state must be a solution of Equation 10 and must satisfy three conditions: *1*) at $X = 0$, $\hat{V} = 0$ because of the voltage clamp, *2*) at $X = L$, $d\hat{V}/dX = 0$ because of the sealed end, and *3*) at $X = A$, $\hat{V} = \hat{V}_A$ must be consistent with the applied current ($\hat{I}_{in}$), which by conservation of current must equal the sum of the core currents flowing away to left and right from $X = A$. This last condition can be expressed

$$\hat{I}_{in} = G_\infty(d\hat{V}/dX)_{X=A^-} + G_\infty(-d\hat{V}/dX)_{X=A^+} \tag{B1}$$

where (because slope is discontinuous at $X = A$) the notations $X = A^-$ and $X = A^+$ designate $X = A$ approached from the left (−) and from the right (+) (see ref. 25 for illustration and explanation). The required AC solution must be expressed differently for the two regions at left and right of $X = A$; thus

$$\hat{V} = \hat{V}_A \frac{\cosh q(L-X)}{\cosh q(L-A)} \quad \text{for} \quad A \leq X \leq L \tag{B2}$$

and

$$\hat{V} = \hat{V}_A \frac{\sinh qX}{\sinh qA} \quad \text{for } 0 \leq X \leq A \tag{B3}$$

These expressions match $\hat{V} = \hat{V}_A$ at $X = A$ and satisfy the boundary conditions at $X = 0$ and $X = L$. Differentiation of Equations B2 and B3 with respect to X and substitution in Equation B1 yields

$$\hat{I}_{in}(A,q) = G_\infty \hat{V}_A \, q[\cosh qA + \tanh q(L - A)] \tag{B4}$$

whereas the AC current detected by the clamp (current flowing from dendrites to clamp at $X = 0$) can be expressed

$$\hat{I}_c(0,q) = -G_\infty(-d\hat{V}/dX)_{X=0} = G_\infty \hat{V}_A \, q/\sinh qA \tag{B5}$$

The ratio of these two currents is independent of the amplitude (implied by $G_\infty \hat{V}_A$) and can be expressed

$$\hat{I}_{in}(A,q)/\hat{I}_c(0,q) = \cosh qA + \sinh qA \tanh q(L - A)$$

$$= \frac{\cosh qL}{\cosh q(L - A)} \tag{B6}$$

where it is not immediately obvious that the last two expressions are equal. To verify this it is helpful to define $B = L = A$ and to note the identities

$$\frac{\cosh qL}{\cosh q(L - A)} = \frac{\cosh q(A + B)}{\cosh qB}$$

$$= \frac{\cosh qA \cosh qB + \sinh qA \sinh qB}{\cosh qB}$$

$$= \cosh qA + \sinh qA \tanh q(L - A)$$

This proves that Equation 24 (where $X = A$ is represented simply as X) holds for any AC frequency and any input location X.

Because Equations B1-B6 hold equally well for the Laplace transform domain, this general result also holds for that domain as asserted with Equations 24 and 29.

REFERENCES

1. Araki, T., and C. A. Terzuolo. Membrane currents in spinal motoneurones associated with the action potential and synaptic activity. *J. Neurophysiol.* 25: 772-789, 1962.
2. Barrett, J. N., and W. E. Crill. Specific membrane properties of cat motoneurones. *J. Physiol. London* 239: 301-324, 1974.
3. Barrett, J. N., and W. E. Crill. Influences of dendritic location and membrane properties on the effectiveness of synapses on cat motoneurones. *J. Physiol. London* 239: 325-345, 1974.
4. Barrett, J. N., and W. E. Crill. Voltage clamp of cat motoneurone somata: properties of the fast inward current. *J. Physiol. London* 304: 231-249, 1980.
5. Brown, T. H., and D. Johnston. Voltage-clamp analysis of mossy fiber synaptic input to hippocampal neurons. *J. Neurophysiol.* 50: 487-507, 1983.
6. Butz, E. G., and J. D. Cowan. Transient potentials in dendritic systems of arbitrary geometry. *Biophys. J.* 14: 661-689, 1974.
7. Calvin, W. H. Dendritic synapses and reversal potentials: theoretical implications of the view from the soma. *Exp. Neurol.* 24: 248-264, 1969.

8. Carnevale, N. T., and D. Johnston. Electrophysiological characterization of remote chemical synapses. *J. Neurophysiol.* 47: 606–621, 1982.
9. Carslaw, H. S., and J. C. Jaeger. *Conduction of Heat in Solids.* London: Oxford Univ. Press, 1959.
10. Finkel, A. S., and S. J. Redman. The synaptic current evoked in cat spinal motoneurones by impulses in single group Ia axons. *J. Physiol. London* 342: 615–632, 1983.
11. Frank, K., M. G. F. Fuortes, and P. G. Nelson. Voltage clamp of motoneuron soma. *Science* 130: 38–39, 1959.
12. Horwitz, B. An analytical method for investigating transient potentials in neurons with branching dendritic trees. *Biophys. J.* 36: 155–192, 1981.
13. Horwitz, B. Unequal diameters and their effects on time varying voltages in branched neurons. *Biophys. J.* 41: 51–66, 1983.
14. Jack, J. J. B., D. Noble, and R. W. Tsien. *Electric Current Flow in Excitable Cells.* London: Oxford Univ. Press, 1975.
15. Jack, J. J. B., and S. J. Redman. An electrical description of the motoneurone and its application to the analysis of synaptic potentials. *J. Physiol. London* 215: 321–352, 1971.
16. Johnston, D., and T. H. Brown. Interpretation of voltage-clamp measurements in hippocampal neurons. *J. Neurophysiol.* 50: 464–486, 1983.
17. Joyner, R. W., J. W. Moore, and F. Ramon. Axon voltage-clamp simulations. III. Postsynaptic region. *Biophys. J.* 15: 37–54, 1975.
18. Perkel, D. H., and B. Mulloney. Electrotonic properties of neurons: steady state compartmental model. *J. Neurophysiol.* 41: 621–639, 1978.
19. Rall, W. Branching dendritic trees and motoneuron membrane resistivity. *Exp. Neurol.* 1: 491–527, 1959.
20. Rall, W. Membrane potential transients and membrane time constant of motoneurons. *Exp. Neurol.* 2: 503–532, 1960.
21. Rall, W. Theory of physiological properties of dendrites. *Ann. NY Acad. Sci.* 96: 1071–1092, 1962.
22. Rall, W. Theoretical significance of dendritic trees for neuronal input-output relations. In: *Neural Theory and Modeling*, edited by R. Reiss. Stanford, CA: Stanford Univ. Press, 1964, p. 73–97.
23. Rall, W. Distinguishing theoretical synaptic potentials computed for different soma-dendritic distributions of synaptic input. *J. Neurophysiol.* 30: 1138–1168, 1967.
24. Rall, W. Time constants and electrotonic length of membrane cylinders and neurons. *Biophys. J.* 9: 1483–1508, 1969.
25. Rall, W. Core conductor theory and cable properties of neurons. In: *Handbook of Physiology. The Nervous System. Cellular Biology of Neurons*, edited by J. M. Brookhart and V. B. Mountcastle. Bethesda, MD: Am. Physiol. Soc., 1977, vol. 1, pt. 1, chapt. 3, p. 39–97.
26. Rall, W., R. E. Burke, T. G. Smith, P. G. Nelson, and K. Frank. Dendritic location of synapses and possible mechanisms for the monosynaptic EPSP in motoneurons. *J. Neurophysiol.* 30: 1169–1193, 1967.
27. Rall, W., and J. Rinzel. Branch input resistance and steady attenuation for input to one branch of a dendritic neuron model. *Biophys. J.* 13: 648–688, 1973.
28. Redman, S. J. The attenuation of passively propagating dendritic potentials in a motoneurone cable model. *J. Physiol. London* 234: 637–664, 1973.
29. Rinzel, J., and W. Rall. Transient response in a dendritic neuron model for current injected at one branch. *Biophys. J.* 14: 759–790, 1974.
30. Selverston, A. I., D. F. Russell, J. P. Miller, and D. G. King. The stomatogastric nervous

system: structure and function of a small neural network. *Prog. Neurobiol.* 7: 215–290, 1976.
31. Turner, D. A., and P. A. Schwartzkroin. Steady-state electrotonic analysis of intracellularly stained hippocampal neurons. *J. Neurophysiol.* 44: 184–199, 1980.

TEN

Comparison of Voltage Clamps With Microelectrode and Sucrose-Gap Techniques

John W. Moore

Department of Physiology, Duke University Medical Center, Durham, North Carolina

Sucrose Gap for Nonmyelinated Axons: Advantages, Problems • **Control of Membrane Potential:** General concepts, Membrane resistance–to–access resistance ratio, Speed of response, Criterion for voltage clamp, Spatial voltage uniformity • **Series Resistance:** Effects and oscillations with compensation, Measurement of series resistance, Compensation for series resistance, Gating currents • **Summary**

This chapter provides a reference background against which the performance of the microelectrode voltage clamps can be appreciated. It is based primarily on personal experience gathered from extensive use of the sucrose-gap technique on lobster and squid axons and is compared with various aspects of this work using microelectrodes.

The sucrose-gap voltage clamp is designed for use with single axons a few millimeters in length—long enough to traverse a chamber in which there are two streams of flowing sucrose gaps separated by the flowing extracellular solution. The axon must extend into side pools that are used for measuring potential and injecting current. The application of sucrose along a length of axon mimics the insulation provided by myelin on myelinated fibers.

The sucrose-gap method of measuring membrane potentials with external electrodes was introduced in 1954 by Robert Stampfli (15), who studied a small bundle of myelinated frog fibers. In 1962 Julian et al. (8) adapted this method to voltage clamp a short segment of the membrane of a single giant lobster axon (9). Soon this method was used to describe the unusually selective blocking of the sodium channel in this axon by tetrodotoxin (14). The method was also extended to studies of the internally perfused squid axons (12).

Sucrose Gap for Nonmyelinated Axons

The experimental convenience of a sucrose gap as developed by Julian et al. (8) for the lobster axon is shown in Figure 1. The *left panel* shows a lobster nerve threaded through holes in two Lucite walls immersed in seawater. In the *right panel* sea water flows from *top* to *bottom* and sucrose flows through small holes in the barriers and out along the axon. The Schlieren boundaries seen in the figure are the changes in the refraction index at the junction of the solutions. The width of this artificial node can be adjusted by controlling the flow rate of the seawater and high-resistance sucrose solution. [Sucrose solutions are made of deionized water and are passed through a filter and a deionizer before use].

In Figure 2 the *upper panel* shows a schematic diagram of a perfused squid axon in a sucrose gap and the solution flow pattern and the *lower panel* shows the electrical equivalent circuit. The membrane potential, dominated by the potassium equilibrium potential, is reduced to nearly zero in the right pool by flowing isosmotic KCl. This treatment increases the potassium conductance so effectively that the membrane is indicated

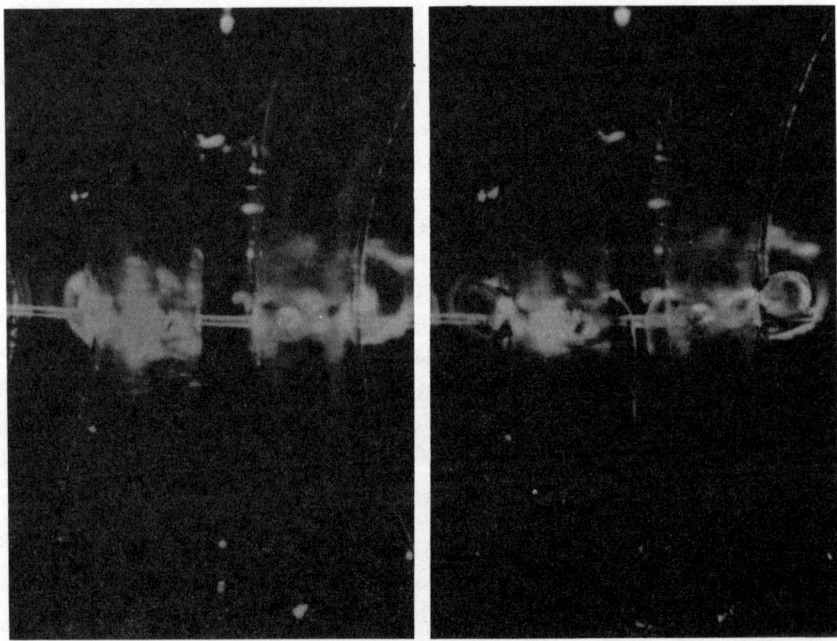

Fig. 1. Lobster nonmyelinated axon threaded through 2 Lucite partitions. *Left*: no sucrose is flowing. *Right*: sucrose is flowing, approaching axon from below and through Lucite partitions. Note open exposed segment or artificial node on axon in center. [From Julian et al. (8).]

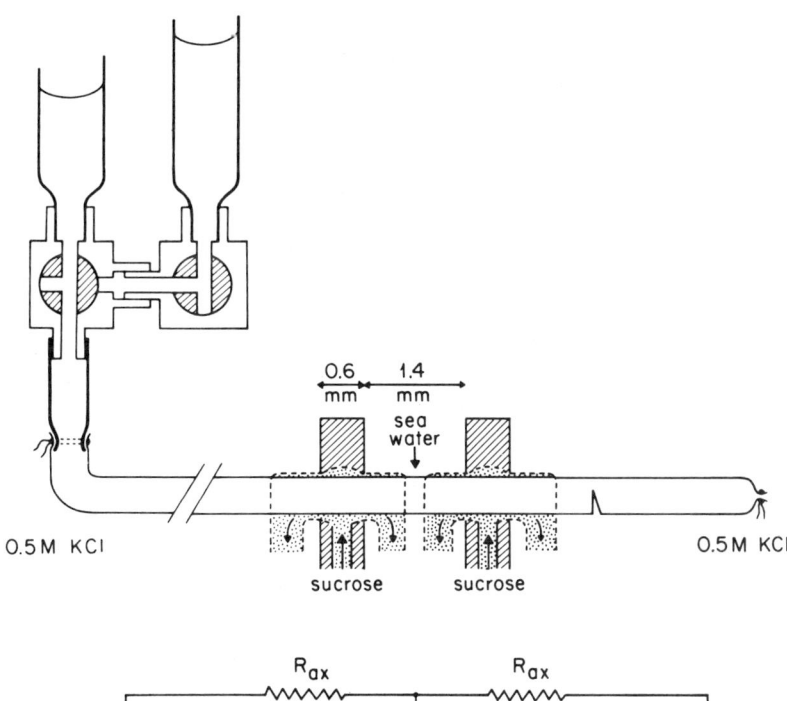
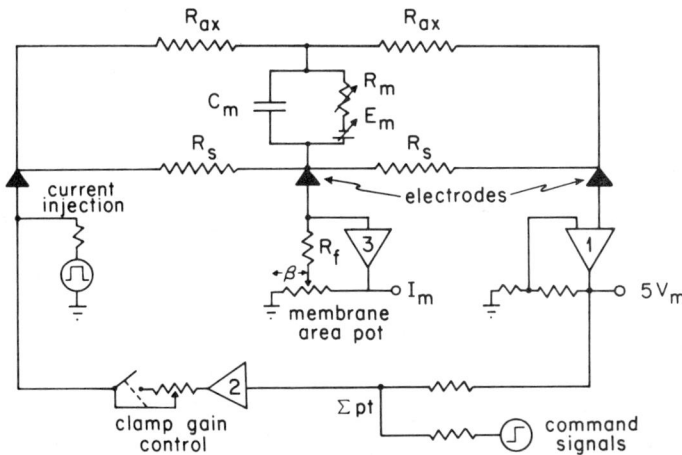

Fig. 2. Experimental arrangement. *Top*: *striped areas*, Lucite partitions; *stippled areas*, sections of axon bathed in flowing isotonic sucrose solution. Axon extends through 600-μm-diam holes in each Lucite partition. Between sucrose solutions, small node of axon is exposed to continuously moving stream of seawater. Solutions in I and V pools also continuously flowing. *Bottom*: electrical equivalent circuit of arrangement in *top*. R_{ax}, resistance through axoplasm; R_s, resistance through sucrose; solution; C_m, R_m, and E_m, membrane capacitance, resistance, and potential of node, respectively; $5V_m$, 5 times potential difference between V pool and virtual ground; I_m, current flowing through node membrane. Current is injected via I pool through isolation resistor. [Adapted from Moore et al. (12).]

as a "short circuit" in the KCl pool. As long as the sucrose insulation is high and the input impedance of the potentiometric amplifier is high compared with both the axial resistance and the resistance across the sucrose stream, the amplifier output will be a faithful measure of the potential inside the membrane at the center node. Very short lengths of axon, usually one-fourth to one-half of the axon diameter, are exposed to the electrolyte solution in the central pool; this minimizes the voltage difference from one edge of the node to the other. It is routinely possible to achieve a resistance of 10–30 MΩ between the central and right pools. Thus, for a nodal-membrane resistance on the order of 50–500 kΩ for squid axons, there is essentially no attenuation of the potential measured between the right pool and ground. This potential very closely approximates that across the nodal membrane because the central pool is held at virtual ground via connection to the input of a current-measuring operational amplifier. The feedback fraction of the current-amplifier output (β) was adjusted to be proportional to the membrane area measured by microscopic observation; this provided an output voltage from the amplifier proportional to the membrane current density. This node may be stimulated to give an action potential when current is injected through a high resistance (Fig. 2, *lower left*). It can be voltage clamped by closing the switch and short-circuiting the resistor connected to the output of the V-clamp amplifier. Then the voltage measured across the node is forced to match a set of command potentials by the voltage control circuit.

Advantages

The sucrose-gap voltage clamp offers several advantages for large nonmyelinated fibers. Whereas most axial-wire techniques require 1–2 cm of axon in good condition, the sucrose-gap technique is very forgiving of inadvertent slips in dissection. It requires good condition only along that length of axon actually spanning the streams of sucrose. Furthermore, when certain experimental conditions destroy a single node, a new one can be obtained simply by moving the axon along through the three streams with a mechanical device that secures both ends; a new patch (length) with good activity can be located very rapidly. A squid axon several centimeters long may provide up to 8–10 fresh nodes; there are usually 3 or 4 of these in a lobster axon. The axon usually is moved enough to avoid making a node from an area that had been surrounded by flowing sucrose because the internal ion content may have been altered in that area; i.e., any ions that left the axoplasm, going through the membrane into the sucrose, would simply have been washed away. Thus the sucrose gap provided the first "patch clamp."

Internal perfusion of axons with experimental solutions is convenient with the sucrose-gap technique, and internal longitudinal flow is not hindered by an internal electrode assembly (Fig. 2). The assembly for the internal perfusate reservoirs can be mounted on the heavy mechanical device that translates the axon.

Problems

The major inconvenience with the sucrose-gap technique results from the leaching out of axoplasmic ions; this increases the axoplasm resistance so that after some time the clamp is no longer really effective. For lobster axons this may occur in 15–30 min. For squid axons with larger diameters the surface-to-volume ratio is much more favorable, and experiments of 1–2 h present no significant problems. The process of ions leaching out with the sucrose stream requires that ions have to move in from other areas to replace them. This causes a minute current flow in the central node resulting in membrane hyperpolarization. The amount of hyperpolarization usually depends on the length of the artificial node; the narrower it is, the greater the hyperpolarization. For voltage-clamp experiments, the hyperpolarization generally presents no problem because it is standard procedure to hyperpolarize the membrane and completely activate the sodium conductance system. There is no appreciable difference in axon performance between those hyperpolarized electrically or by this method. Thus, whereas the sucrose-gap method is not useful in measuring resting potentials, it is quite appropriate for voltage clamping.

Another inconvenience of the sucrose-gap method is that the flow of the streams is not completely controllable. Tiny bits of dirt can cause alterations in the flow pattern, changing the area of the node. This problem is circumvented by measuring the area, proportional to the capacitance of the nodal membrane, when a test pulse is not being applied; i.e., the axon is "time-shared" for the two measurements—capacitance between pulses (1–2 s) and voltage-clamp currents during the few milliseconds when the pulse is on. The nodal membrane capacity is measured by its response to a 10- to 20-kHz sine wave and is used to automatically set the area potentiometer so that the output is always proportional to current density, regardless of flow pattern alterations.

Control of Membrane Potential

General concepts

To appreciate some of the differences between sucrose gaps and microelectrodes, consider the general problems of voltage clamping. A generalized circuit for voltage control is shown in Figure 3. The voltage (V_m)

222   J. W. Moore

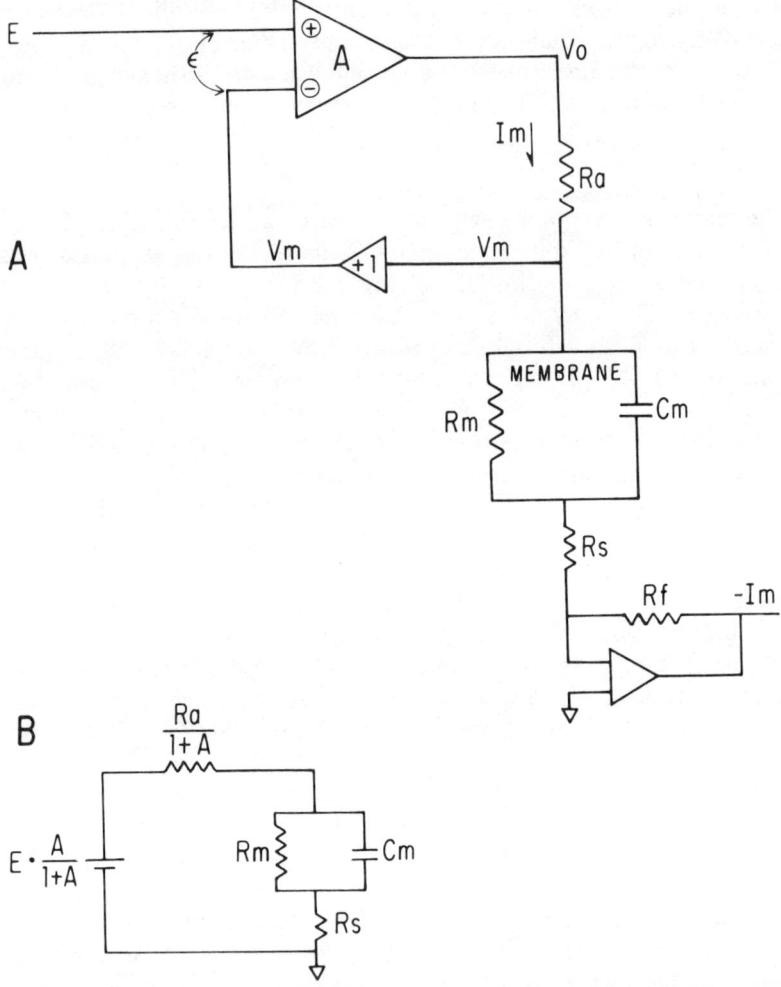

Fig. 3. *A*: simplified diagram of generalized voltage clamp of membrane. *B*: effective equivalent circuit of voltage clamp.

between the membrane and ground is measured by a potentiometric or electrometer amplifier (amplifier that draws negligible current) with a gain of +1. This amplifier's output is compared with a set of input command signals at the summing point of an operational amplifier (10). Any small potential difference (ϵ) between the command signal ($-E$) and the potential (V_m) is amplified by a large gain (A) to give an output voltage (V_o) for potential control. This output voltage is applied to the membrane via an access resistance (R_a) with a polarity that makes the potential across the membrane equal the negative sum of the command potentials.

There is always an access resistance between the output of the control amplifier and the interior surface of the membrane. It may take the form of a microelectrode or of the resistance of a metal to an electrolyte junction. It will also include axoplasmic or cytoplasmic resistance. The relative contributions of electrodes and cytoplasm will vary with each preparation and with each cell type and size but will always have to be considered because they are important aspects in determining the speed and accuracy of the control circuit.

Also included is a small resistance in series with the membrane that is important (at least for squid axons). The membrane current (I_m) is measured by an operational amplifier with the currentometric feedback configuration, which maintains its input at or near ground potential. The membrane current flows through the feedback resistor (R_f) of the current-to-voltage converter, producing a voltage proportional to $I_m R_f$.

The equations that describe the operation of this control circuit can be developed very simply. The output of the control amplifier is equal to the difference in the input potentials ($E - V_m$) times the gain of amplifier A

$$V_o = A(E - V_m) \tag{1}$$

where V_m is the membrane potential. [Note that this is true because E is connected to the positive input and V_m is connected to the negative input.] The output potential is distributed between the drop across the access resistance and the membrane potential or

$$V_o = V_m + R_a I_m \tag{2}$$

Solving these two equations for the potential across the membrane gives

$$V_m = EA/(1 + A) - R_a I_m/(1 + A) \tag{3}$$

This description applies to the simple circuit shown in Figure 3B, which is called an equivalent circuit for the voltage clamp. The figure shows that the voltage control circuit is equivalent to a battery, in series with a resistance, that drives the element being controlled, in this case an active membrane. Here the potential source equals $EA/(1 + A)$. For large values of A, the battery potential approaches E very closely. However, if A is not large (e.g., at high frequencies), the difference between the command potential and the effective battery driving the membrane can be substantial.

The equivalent resistance in series with the battery is R_a divided by the factor $1 + A$. A large amplifier gain reduces the effect of the access resistance to a small value, but a low gain degrades the overall performance of the circuit from ideal because of the IR_a voltage drop.

Thus the element being controlled is effectively put across the terminals of a battery with a low internal resistance. As the gain of the amplifier increases, the value of the battery approaches the command potential, and the effect of the series access resistance is minimized; the potential across the load [membrane plus series resistance (R_s)] very nearly approaches the command potential and an ideal voltage control is obtained. A small series resistance is closely associated with the membrane. The feedback potential is measured across it plus the membrane and its effect is not reduced by the feedback gain. The access resistance is said to be "inside" the negative-feedback circuit, whereas the series resistance is said to be "outside" the feedback loop. For the squid membrane, R_s is so closely associated with the membrane that it is extremely difficult to penetrate between it and the membrane, even with the finest microelectrode. This resistance is included for purposes of generality even though it may not seem to be important for all excitable membranes. If a potential probe can be placed between R_s and the membrane, the spurious IR_s voltage drop can be subtracted from V_m and a better membrane voltage clamp obtained.

Membrane resistance–to–access resistance ratio

The fraction of the control output voltage that is fed back to the control amplifier is $R_m/(R_m + R_a)$, where R_m is the membrane resistance. If R_m is not large compared with R_a, the effective loop gain of the control circuit may be so reduced that the voltage-clamp performance is degraded.

For the sucrose gap this access resistance is largely the axial or longitudinal resistance through the axoplasm, which is surrounded by the sucrose, but it also includes the relatively low impedance of a large bath electrode. For a squid axon (500-μm diam) the value of R_a will range from ~1 kΩ in a fresh fiber to 10 kΩ in an axon that has been in flowing sucrose for an hour or so and had its internal ions leached out (see *Problems*, p. 221). The patch of membrane being clamped may be only 0.002 cm^2 and have a resting resistance of 0.5–1 MΩ. Thus the ratio of resting membrane to access resistance may be as high as 10^3. In the active state the membrane resistance may fall 100-fold and lower the ratio to 10. Even in this "worst-case" condition, 90% of the output signal is fed back.

However, for a micropipette electrode the access resistance will be several tens of megohms and the input resistance of the cell being clamped may range from 1 to 100 MΩ. Thus R_m/R_a will frequently be on the order of 1. During activity a 100-fold decrease in membrane resistance will reduce this ratio to 0.01. Therefore it becomes clear that the reduction of the control-circuit loop gain by the R_a-R_m voltage divider, which is not of much concern for the sucrose gap, may make adequate clamping via micropipettes impossible. This problem is exacerbated in single-electrode voltage clamps, which time share the pipette for current injection and potential measurement. The actual current injected must be twice the average value; the voltage drop across the electrode is doubled when passing current, and thus the effective value of the access resistance becomes twice the value of R_a.

Speed of response

One measure of the quality of a voltage clamp is the reduction of the time constant for changing the membrane potential from that observed under current control. This can be seen in Figure 3B in which the membrane-charging time constant is $C_m R_p/(1 + A)$ [where $R_p = R_a R_m/(R_a + R_m)$] in the voltage clamp but $C_m R_m$ for the current clamp. The $R_a C_m$ product controls the maximum speed of response under voltage clamp. Thus it is clear why the sucrose-gap voltage clamp can be so much faster than a micropipette circuit.

Criterion for voltage clamp

In 1960 Cole and Moore (4) set forth a formal and obvious criterion "the potential difference across the membrane capacity shall have a known and constant value during a time and over the area of the membrane in which the current flow is measured." They also said that the

> criterion can probably be met only by a constant potential applied by perfect electrodes in direct contact with the membrane capacity. Until such an ideal experiment is possible, the effects of practical concessions to axon survival and equipment capabilities upon the experimental results need to be examined.

Spatial voltage uniformity

In contrast to the internal axial-wire voltage clamp on the squid giant axon, which depends on the low surface resistance of the axial wire, the sucrose gap simply limits the longitudinal voltage gradient by restricting the width of the gap to a small fraction (¼–½) of the diameter. This gives a clean record without notches, as shown in the original work by Julian et al. (8). This has also been examined by a computer simulation in which the axon was represented by the cable equations and the membrane endowed with Hodgkin-Huxley activity (13). When the length of the node equaled the axon diameter, the maximum voltage deviation of the membrane from the point of current injection to the point of potential measurement was ~20% of the command pulse. The error was far less when the ratio of the length of the node to the diameter was 0.5 or less. Thus we are assured that with this criterion we have excellent voltage uniformity. Generally micropipettes are used to control the voltage at a certain point or to "point clamp" a cell. The membrane and cable properties of the cell determine the spatial voltage profiles. The lack of spatial voltage uniformity will compromise the accuracy and quality of current, conductance, and reversal-potential records unless the biological current source is restricted to a small region near the controlled point. The significant question is how much will it be compromised. Measurements of conductance can be restricted to the region of membrane near the point clamp by the addition of a high-frequency perturbation to the step depolarizations. The high-frequency conductance can be obtained with a Fast Fourier Transform from current and voltage records at the control point (M. Hines and J. W. Moore, unpublished observations; see also ref. 7).

Series Resistance

Effects and oscillations with compensation

The detection or measurement of series resistance is difficult with microelectrodes because of the limited bandwidth of current injection and the cable properties of cells. Nevertheless it is so important for measurement of ionic conductances with axial-wire and sucrose-gap voltage clamps of axon membranes that a brief discussion of its consequences and compensation are given.

The squid giant axon is sheathed by a layer of Schwann cells (1, 5). Membrane currents in the axon traverse the clefts between the cells and produce a voltage drop across this resistance in series with the membrane. This voltage drop causes the measured internal potential to deviate from the potential actually across the active membrane. With a voltage clamp this voltage error gives rise to changes in the shape of the ionic current (16); these deviations are not tolerable when developing a precision description of and model for the ionic conductance systems. Therefore, for precision measurement of kinetics and for distinguishing between models, it is essential that the series resistance be accurately known and fully compensated.

The major problem with compensation for the IR_s voltage error introduced by voltage clamping excitable membranes is that oscillations begin to appear (or increase in amplitude) throughout the circuit as the error correcting signal (proportional to membrane current) is increased toward full compensation. In their pioneering work on the voltage clamp, Hodgkin et al. (6) noted this problem and usually settled for ~60% of full compensation—a compromise between clamp stability and the desired full compensation for their observed value of 7.0 Ωcm^2. This residual uncompensated R_s of 2.8 Ωcm^2, for their peak sodium currents of ~1 mA/cm^2, gave a maximum error of ~3 mV. However, for the much larger peak sodium currents (5–7 mA/cm^2) seen by Cole and Moore (4), such an uncompensated residual R_s would have given an intolerable error of nearly 20 mV. Noting the contributions of electrode, axoplasm, and bathing solution to the total series resistance, they placed microelectrodes just inside and outside the membrane to monitor its potential difference as closely as possible (see Figs. 2 and 3 of ref. 4). This put much of the series resistance within the feedback loop and eliminated its effect. Cole and Moore (4) measured a residual R_s of 2 Ωcm^2 and were able to compensate for at least two-thirds of this, which resulted in a net uncompensated R_s of <1 Ωcm^2.

Nevertheless many investigators of squid axons still employ a potential-measurement system similar to that used by Hodgkin et al. (6); they must also deal with a residual series resistance similar to that observed by Hodgkin et al. (6) (~7 Ωcm^2) with intact axons. When axons are internally perfused with solutions of low ionic strength, considerably larger values of series resistance are encountered. To avoid large series-resistance errors, several investigators reduced the sodium current by bathing the axon in a solution with a low sodium concentration.

Measurement of series resistance

The usual technique used in my laboratory for determining the series resistance was developed by Moore and Cole (11), who measured the voltage jump in an axon in response to a current step. The requirements for precise measurement with this widely used method have been carefully analyzed (unpublished observations and ref. 2). A 1-μs lag (time constant) in the current reduced the measured value of the series resistance by 1 Ωcm^2. Similar lags in the voltage response system give similar and further reductions in the observed value of series resistance. Therefore it has been standard practice in my laboratory to include a known resistance in series with the passive RC model for the membrane and to test the ability of the system to measure this resistance accurately. The system can detect the known resistance with a resolution of 0.5 Ωcm^2, establishing the validity of the R_s measurements that range from 1.0 to 2.5 Ωcm^2.

Compensation for series resistance

I use an alternative method, developed below, for full compensation of ionic current errors caused by series resistance without introduction of

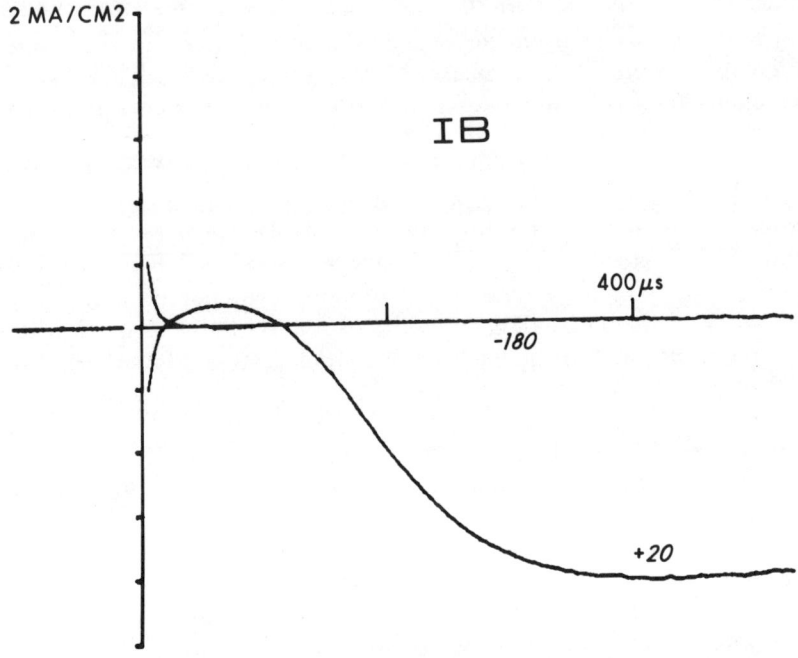

Fig. 4. Membrane current signal (*IB*) after passing through active electronic bridge. Current resulting from brief hyperpolarization to −180 mV is balanced to brief transient (*upward*, 25 μs) before almost flat trace. With this bridge balance, when current associated with strong depolarizing voltage step to +20 mV has linear capacitance components and leakage subtracted by bridge, gating current (*upward arc*) can be seen to precede the characteristic ionic current carried by Na^+.

oscillations. The requirement of full compensation for the series resistance means that the membrane capacitance must be charged in an infinitesimal time by an infinite current. This cannot be physically realized. Not only do practical amplifiers have limited bandwidth and current output capabilities, but the delay between the input and output (exacerbated by capacitive load of membrane) means that the compensating signal is delayed relative to the time needed. This delay in the feedback signal to compensate for the voltage error in charging the capacitance is the source of the oscillations in the control of the membrane voltage. In other words the resistance in series with the membrane capacitance is necessary (and often sufficient) for stability against oscillations.

Recognizing this, the capacitive current errors were accepted to obtain accurate ionic current records. The capacitive current is separated and only the ionic currents for R_s compensation are fed back. This technique allows one to achieve not only full compensation for ionic currents but

Table 1
Comparison of microelectrode and sucrose-gap voltage-clamp techniques

	Microelectrode	Sucrose Gap
Appropriate geometry	Any cell of reasonable size	Axon
Accessibility	Tip penetration only, dissection often necessary	Axon dissected and cleaned
Voltage uniformity	Rarely good, geometry and channel-density dependent	Excellent, 90%–95%
Multiple studies	Penetration of new cell	Slide along axon
Flowing solution required	Only to change medium	Large volume, electrically isolated
Capacitance from control amp to V input	Significant, shielding necessary	Negligible
τ of V_m measurement	>25–100 μs	<3 μs
R_a/R_m		
Resting	0.1–10	~10^{-3}
Active	10–1,000	~10^{-1}
R_s/R_m		
Resting	Unknown	10^{-3}
Active	Unknown	10^{-1}
Voltage error across R_s	Maximum $R_s I_m$	15 mV
R_s compensation	Rarely used	Full

V, voltage; τ, time constant; V_m, membrane voltage; R_a, access resistance; R_s, series resistance; R_m, membrane resistance.

also even overcompensation for the error of series resistance without driving the system into oscillations.

The method was conveniently implemented because an active (electronic) bridge circuit is normally used to balance out the leakage current and linear components of the capacitive currents. In squid axons the latter contain not a single but an infinite series of time constants [as shown by Cole and Cole (3)]. This infinite series can be represented and balanced out reasonably well by summing three exponential decays of different time constants and amplitudes.

Gating currents

With such a fast and stable clamp, in which the linear component of the membrane capacitance is electronically balanced out along with the leakage currents (for hyperpolarizing voltage pulse), it is possible to observe asymmetrical gating currents in an individual record without averaging. Figure 4 shows the bridge output (*IB*) balanced for a hyperpolarizing pulse to −180 mV and the early outward (upward) gating current followed by the inward sodium current for a depolarization to +20 mV.

Summary

The sucrose-gap voltage clamp demonstrates a standard for comparison in terms of speed, voltage uniformity, and compensation for series resistance. The reasons why micropipette voltage clamps are necessarily much slower are discussed and a comparison of some of the ways in which the two techniques can be compared are given in Table 1.

REFERENCES

1. Adelman, W. J., Jr., J. Moses, and R. V. Rice. An anatomical basis for the resistance and capacitance in series with the excitable membrane of the squid giant axon. *J. Neurocytol.* 6: 621–646, 1977.
2. Binstock, L., W. J. Adelman, Jr., J. P. Senft, and H. Lecar. Determination of the resistance in series with the membrane of giant axons. *J. Membr. Biol.* 21: 25–47, 1975.
3. Cole, K. S., and R. H. Cole. Dispersion and absorption in dielectrics. I. Alternating current characteristics. *J. Chem. Phys.* 9: 341–351, 1941.
4. Cole, K. S., and J. W. Moore. Ionic current measurements in the squid giant axon membrane. *J. Gen. Physiol.* 44: 123–167, 1960.
5. Geren, B. B., and F. O. Schmitt. The structure of the Schwann cell and its relation to the axon in certain invertebrate nerve fibers. *Proc. Natl. Acad. Sci. USA* 40: 863–871, 1954.
6. Hodgkin, A. L., A. F. Huxley, and B. Katz. Measurement of current-voltage relations in the membrane of the giant axon of *Loligo*. *J. Physiol. London* 116: 424–448, 1952.
7. Johnston, D., and T. H. Brown. Interpretation of voltage-clamp measurements in hippocampal neurons. *J. Neurophysiol.* 50: 464–507, 1983.
8. Julian, F. J., J. W. Moore, and D. E. Goldman. Membrane potentials of the lobster giant axon obtained by use of the sucrose-gap technique. *J. Gen. Physiol.* 45: 1195–1216, 1962.
9. Julian, F. J., J. W. Moore, and D. E. Goldman. Current-voltage relations in the lobster giant membrane under voltage clamp conditions. *J. Gen. Physiol.* 45: 1217–1238, 1962.
10. Moore, J. W. Operational amplifiers. In: *Physical Techniques in Biological Research*. New York: Academic, 1963, p. 77–97.
11. Moore, J. W., and K. S. Cole. Voltage clamp techniques. In: *Physical Techniques in Biological Research*. New York: Academic, 1963, p. 263–321.
12. Moore, J. W., T. Narahashi, and W. Ulbricht. Sodium conductance shift in an axon internally perfused with a sucrose and low-potassium solution. *J. Physiol. London* 172: 163–173, 1964.
13. Moore, J. W., F. Ramon, and R. Joyner. Axon voltage-clamp simulations. I. Methods and tests. *Biophys. J.* 15: 11–24, 1975.
14. Narahashi, T., J. W. Moore, and W. R. Scott. Tetrodotoxin blockage of sodium conductance increase in lobster giant axons. *J. Gen. Physiol.* 47: 965–974, 1964.
15. Stampfli, R. A new method for measuring membrane potentials with external electrodes. *Experientia* 10: 508, 1954.
16. Taylor, R. E., J. W. Moore, and K. S. Cole. Analysis of certain errors in squid axon voltage clamp measurements. *Biophys. J.* 1: 161–202, 1960.

ELEVEN

Voltage Clamping Small Cells

Harold Lecar
Laboratory of Biophysics, National Institute of Neurological and Communicative Disorders and Stroke, Bethesda, Maryland

Thomas G. Smith, Jr.
Laboratory of Neurophysiology, National Institute of Neurological and Communicative Disorders and Stroke, Bethesda, Maryland

---

Speed of Response of Two-Microelectrode Clamp: The clamp as a linear feedback network, Wide-band clamp with different membrane elements, Limitations because of amplifier bandwidth, Two–time-constant control amplifier, Improving clamp response with current feedback, How to realize fast rise time • **Voltage Clamping Tissue-Cultured Neurons** • **Space-Clamp Considerations:** Short cylindrical cell, Spherical soma, Pancake-shaped cell or syncitial sheet, Cells with dendritic networks • **Whole-Cell Patch Clamp:** Paradigm for use of current feedback in single-electrode arrangement • **Conclusion**

In this chapter, we review some basic considerations confronting an experimenter starting out to clamp a new preparation. Is it preferable to use one or two microelectrodes or to try the new patch-electrode techniques? What ultimately limits the speed of response of a microelectrode clamp? When will cell geometry preclude accurate spatial voltage control? Some of these questions have been considered in greater detail in previous chapters, and our aim in this chapter is to give the reader a qualitative feel for these matters. In addition, we discuss some of the newer less-established techniques of clamping and deal with a few unusual special-case situations. As will become apparent, these techniques and situations arise mainly when clamping small (<20 µm diam) cells. The whole-cell patch clamp, for example, takes advantage of the otherwise difficult properties of small cells and employs novel "tricks" to achieve the desired result of fast voltage clamping.

Until recently the voltage clamp of small cells was problematic. Two-microelectrode voltage clamp requires the insertion of two high-impedance electrodes into the cell without excessive damage. The high-impedance microelectrodes impose experimental design problems—the slow response of the voltage sensor must be compensated, the electrodes may be coupled to each other and to ground by stray capacitance, and the clamp must have sufficient gain and deliver sufficient current to overcome the high access resistance of the current-carrying microelectrode. The design problems can be divided into two classes: those involved in optimal design of the electronic feedback circuit, which are discussed in the chapter by Finkel and Gage, and those that emanate from the electrical characteristics of the preparation, even when the electronics function ideally. These latter considerations are the focus of the next section.

An alternative type of clamp, discussed in the chapter by Finkel and Redman, is the intermittent single-microelectrode clamp. This method is most applicable for clamping cells imbedded in tissue where the cell must be impaled "blind." The design of such clamps is limited by the speed with which the circuitry can be switched, and generally one does not expect response times comparable with those for the two-microelectode clamp.

A third alternative is now available for tissue-cultured cells and other cells that can be directly visualized. This is the so-called whole-cell patch clamp, which is discussed in this chapter. The use of patch-clamp electrodes instead of microelectrodes allows a low-impedance access to a small cell. This in turn allows an approximate voltage clamp to be obtained with a single microelectrode. If the electrode is sufficiently low impedance and the cell is sufficiently small the errors are not large and the response time is quite rapid.

Speed of Response of Two-Microelectrode Clamp

The purpose of a two-electrode voltage clamp is to control the membrane potential by using the deviation between recorded and desired membrane potential as the error signal in a feedback loop. In the simplest representation (Fig. 1), the control amplifier of high gain $[K(s)]$ amplifies the difference between the membrane potential (V_m) and the command signal (V_c). Here, as in the chapter by Finkel and Gage, the variables are taken to be Laplace transforms; thus K, V_m, etc. are all functions of the transform variable s, which has the dimensions of frequency. The output voltage (V_0) is dropped across the membrane in series with the current-delivering microelectrode. A perfect voltage clamp can deliver a step of potential V_m regardless of changes in electrode or membrane impedance. In Figure 1 the clamp output potential is

$$V_0 = K(V_c - V_m) \tag{1}$$

and the membrane potential is

$$V_m = V_0 - I_m R_e \tag{2}$$

where I_m is the membrane current and R_e is the current-electrode resistance. Combining these equations to eliminate V_0

$$V_m = KV_c/(K + 1) - I_m R_e/(K + 1) \qquad (3)$$

When K is very large, the electrode access resistance is effectively reduced to R_e/K and the membrane potential approaches V_c. The actual performance of the feedback system requires specification of the frequency dependence of $K(s)$ and of the membrane admittance. A realistic model of the clamp would also contain the explicit frequency dependence of the microelectrodes, of their coupling by stray capacitance, and of other electronic elements used for compensation. These considerations can be quite complex (see the chapter by Finkel and Gage and refs. 1 and 12).

The clamp as a linear feedback network

There are two criteria for adequate clamp performance: *1*) that a step of potential be delivered to the membrane in a time (rise time or settling time) that is short compared with the relaxation time of membrane excitation processes and *2*) that, after the step is established, the system is still stable in the presence of excitable elements.

For most purposes, the membrane and electrodes can be represented by an equivalent linear admittance so that V_m and V_c can be redefined relative to some appropriate potential (such as resting potential). Thus

$$I_m(s) = Y_m(s) V_m(s) \qquad (4)$$

where Y_m is either the admittance of the resting membrane or the effective linearized admittance for an active membrane. Here we recast

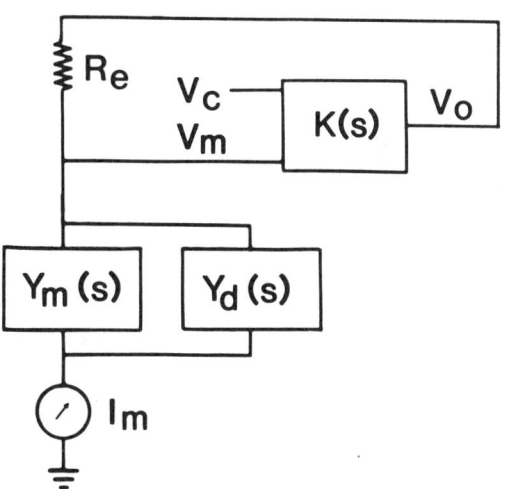

Fig. 1. Simplified schematic diagram of two-microelectrode voltage clamp. Membrane potential (V_m) is assumed to be recorded with perfect fidelity; filtering by microelectrode capacitance is perfectly compensated in this simplified picture. Resting somatic membrane admittance [$Y_m(s)$] is shown in parallel with admittance of passive dendritic network [$Y_d(s)$]. V_c, command signal; $K(s)$, voltage-clamping system with output $V_0 = K(V_c - V_m)$; R_e, current-carrying microelectrode resistance; I_m, membrane current.

Equation 3 in terms of Laplace transforms so that V_m is $V_m(s) = \mathscr{L}[V_m(t)]$ and so forth. With this change and with Equation 4 substituted into Equation 3

$$V_m = [K/(K + 1)]V_c - [Y_m R_e/(K + 1)]V_m \qquad (5)$$

which can be rearranged to give the transfer function

$$T(s) = V_m/V_c = K/(K + 1 + R_e Y_m) \qquad (6a)$$

Because $K \gg 1$ at frequencies of interest, a convenient form for further discussion is the approximation

$$T(s) = [1 + R_e Y_m(s)/K(s)]^{-1} \qquad (6b)$$

In subsequent discussions we are concerned with different choices of $Y_m(s)$ or $K(s)$ and their effects on the system response.

Wide-band clamp with different membrane elements

To have an idea of the intrinsic limitations on the speed of response of a clamp, we consider Equation 6b with three different choices of $Y_m(s)$: a resting membrane considered as an RC parallel circuit (where R is resistance and C is capacitance), a resting cell whose input admittance is dominated by dendritic cables, and an active membrane with an electrically excitable element. At first, $K(s)$ is assumed to be infinitely wide band. We also estimate the minimum bandwidth needed to approach the idealized response times of the wide-band approximation. For the resting membrane

$$Y_m = G_m + C_m s = G_m(1 + s\tau_m) \qquad (7)$$

where G_m is the membrane conductance, C_m is the membrane capacitance, and τ_m is the membrane time constant $R_m C_m = G_m^{-1} C_m$. A parallel passive dendritic network might have a cable impedance of the form

$$Y_d = G_d(1 + s\tau_m)^{1/2} \qquad (8)$$

where G_d is the equivalent input conductance of the dendrites. Note the square root in Equation 8; this is because dendritic loads have a characteristic admittance that varies more slowly with frequency than a simple RC network.

For an active membrane we could consider the effective linear admittance at the operating point. Thus an electrically excitable membrane obeying the Hodgkin-Huxley equations would have an admittance of the form (4)

$$Y_{HH}(s) = g_\infty + sC_m + \sum_{j=1}^{3} g_j/(1 + s\tau_j) \qquad (9)$$

where g_∞ is the membrane conductance and g_j and τ_j are parameters defined in terms of the dynamics of the excitable elements. In this linearized representation, relaxing voltage-dependent conductances give rise to inductive elements with coefficients g_j, which can be positive or negative depending on the reversal potential of the particular gated pathway.

The simplest example to consider is the response of the resting membrane to a voltage step. Substitution of Y_m from Equation 7 into Equation 6b gives

$$T(s) = [1 + \theta_m(1 + \tau_m s)]^{-1} \tag{10}$$

where $\theta_m = R_e G_m/K_0$ and K_0 is the amplifier gain at zero frequency. A measure of the static error in voltage control, θ_m is a useful dimensionless parameter (see Eq. 12). Essentially, a small θ_m means that a voltage-clamp experiment on a given preparation is feasible. The membrane-potential response to a step function is given by

$$V_m(t) = \mathscr{L}^{-1}[T(s) V_c(s)]$$

which is

$$V_m(t) = V_c \frac{1 - \exp(-t/\tau_R)}{1 + \theta_m} \tag{11}$$

where $\tau_R = R_e C_m/K_0(1 + \theta_m) \cong \tau_e/K_0$, t is time, and τ_e is the electrode time constant. Here the time constant τ_R is a measure of the limiting time needed to charge the membrane capacitance through a microelectrode of a given resistance using a perfect wide-band control amplifier of a specified gain. The factor θ_m is related to the static error by

$$\Delta V = V_c - V_m = \frac{\theta_m V_c}{1 + \theta_m} \tag{12a}$$

so that for $\theta_m \ll 1$

$$\Delta V/V_c \cong \theta_m$$

To assess the feasibility of clamping a particular cell with given fidelity and rise time, we estimate θ_m and τ_R from experimentally determined electrical properties of the system and the preparation. For a 20-μm-diam neuron with 1 μF/cm^2 of capacitance impaled by a 100-MΩ electrode, $\tau_e = 1.3$ ms. Because K_0 is usually of the order of 10^3–10^4, the limiting rise time is of the order of microseconds. In the next section we calculate the actual bandwidth needed to achieve this rise time. Since we are comparing rise times for transients of different functional form, let us define a standard rise time (t_r) as the time to reach 90% of the steady state. Thus for the wide-band example

$$t_r = 2.3\tau_R \tag{12b}$$

If a small cell has an extensive dendritic network, Y_m can be replaced by the cable admittance given in Equation 8. The transfer function becomes approximately

$$T(s) = [1 + \theta_d(1 + s\tau_m)^{1/2}]^{-1} \tag{13a}$$

where θ_d is given by $R_e G_d/K_0$ in analogy to the membrane parameter defined in Equation 11. For this case the response to a depolarizing step can again be obtained by inverting the transform

$$V_m(s) = T(s)V_c(s)$$

which after some algebra becomes

$$V_m(t) = V_c^* \tau_d^{-1/2} \int_0^t [\exp(-t/\tau_d^*)]\{(\pi t)^{-1/2} \tag{13b}$$
$$- \tau_d^{-1/2} \exp(t/\tau_d) \operatorname{erfc}[(t/\tau_d)^{1/2}]\}dt$$

where $\tau_d = \theta_d \tau_m$, $\tau_d^* = \tau_d/(1 + \theta_d)$, and $V_c^* = V_c/(1 + \theta_d)$. Thus under certain circumstances the response of the clamp can be affected by the load even when the electronics are nominally "perfect."

For a membrane with a single electrically excitable conductance, the infinite-bandwidth clamp transfer function becomes second order. When an admittance similar to that in Equation 9 is substituted into Equation 6b

$$T(s) = [1 + \theta_m(s\tau_m + 1)\theta_j(\tau_{ex} + 1)]^{-1} \tag{14}$$

where $\theta_j = R_e g_j/K_0$ and τ_{ex} is the relaxation time constant of the excitable conductance. In this case a negative-resistance term for θ_j can cause the membrane response to be oscillatory. However, for $K \gg 1$ Equation 14 reduces to a first-order response with the basic time constant, τ_R.

Limitations because of amplifier bandwidth

In this section we consider some simple cases in which the clamp bandwidth also comes into play. The amplifier can be considered to have a simple low-pass response

$$K(s) = K_0/(1 + s\tau_i) \tag{15}$$

where τ_i is the time constant of the control amplifier [$\tau_i = (1/2\pi)$(bandwidth)]. Substituting from Equation 15 for $K(s)$ and from Equation 7 for $Y(s)$ into Equation 5, we obtain the transfer function for the system as analyzed by Katz and Schwartz (9). The approximate transfer function is

$$T(s) = 1 + [\theta_m(1 + s\tau_m)(1 + s\tau_i)]^{-1} \quad (16)$$

which can be rewritten as

$$T(s) = (\tau_m \tau_i)^{-1}(s^2 + 2\omega_0 \eta s + \omega_0^2)^{-1} \quad (17)$$

where ω_0 is the resonance frequency and η is damping ratio, a measure of the rapidity with which transient oscillations subside. From Equation 16, ω_0 and η can be written explicitly as

$$\omega_0 = (K_0/\tau_i \tau_e)^{1/2} \quad (18a)$$

$$\eta = (\tau_i \tau_e / 4K_0)^{-1/2}(\tau_e^{-1} + \tau_m^{-1} + \tau_i^{-1}) \quad (18b)$$

The expression for η can be further simplified by defining a load time constant

$$\tau_L = \frac{\tau_e \tau_m}{\tau_e + \tau_m}$$

and letting $\beta = \tau_m/(\tau_e + \tau_m)$ (see ref. 9). Then

$$\eta = (4K_0\beta)^{-1/2}[(\tau_i/\tau_L)^{1/2} + (\tau_L/\tau_i)^{1/2}] \quad (19)$$

The response of this second-order system to a step of voltage is given by

$$V_m(t) = V_c'[1 - f(t)\exp(-\eta\omega_0 t)] \quad (20a)$$

where $V_c' = V_c/(1 + \theta_m)$ and $f(t)$ takes different forms depending on the value of η. Thus the characteristic underdamped [$f_u(t)$], critically damped [$f_c(t)$], and overdamped [$f_o(t)$] responses are given by

$$f_u(t) = \sin(\omega t + \cos^{-1}\eta) \qquad \eta < 1 \quad (20b)$$

$$f_c(t) = 1 + \omega_0 t \qquad \eta = 1 \quad (20c)$$

$$f_o(t) = \sinh(\omega t + \cosh^{-1}\eta) \qquad \eta < 1 \quad (20d)$$

where $\omega = \omega_0(|\eta^2 - 1|)^{1/2}$.

To understand how the parameters of the circuit affect the performance of the system, we must see how η and ω_0 vary as functions of τ_i, τ_e, and τ_m. The damping ratio and resonance frequency are written in terms of the network parameters in Equation 18.

The damping ratio is plotted as a function of control-amplifier response time in Figure 2. The optimum response is often taken to be $\eta = 1$ (i.e., critical damping), although a somewhat faster settling steplike response can be had for η as low as 0.6.

The figure shows that there are two conditions of critical damping and a broad region of response times for which the system response is

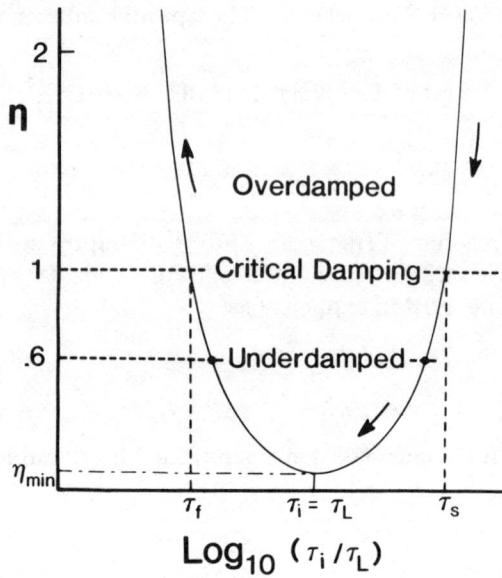

Fig. 2. Generalized plot of damping ratio (η) for second-order approximation of voltage-clamp system as function of instrumentation response time (τ_i) and input load time constant (τ_L). Logarithm of ratio is plotted on abscissa to span large time range. Two conditions of critical damping ($\eta = 1$) are shown as τ_f (fast) and τ_s (slow). Overdamped responses are on curve above and underdamped responses are on curve below critically damped points. Useful ranges of curve are above $\eta = 0.6$. Below 0.6, responses show excessive oscillations in response and long settling times. Ideal is to work above 0.6 on fast limb of curve. At lowest point of curve (η_{min}), system is highly unstable. In general, the further left the locus of curve or operating point of curve, the faster the response time to step function.

underdamped. Thus to have a nonoscillatory response the system must be operated outside of the range of response times between τ_s and τ_f (see Eq. 21). The explicit values of τ_s and τ_f can be obtained by setting $\eta = 1$ in Equation 19. Because $K_0 \gg 1$, they can be approximated as (9)

$$\tau_i \text{ (fast)} \equiv \tau_f = \tau_e/4K_0 = \tau_L/4K_0\beta$$
$$\tau_i \text{ (slow)} \equiv \tau_s = 4K_0\beta\tau_L \tag{21}$$

For the fast condition of critical damping, the rise time is given by

$$t_r = 7.78\tau_L/4K \tag{22}$$

Comparing this with the infinite-bandwidth rise time of Equation 12b, we see that very little improvement is had beyond the condition of fast critical damping. Figure 3 shows how the damping ratio curve changes with amplifier gain. Higher gain pushes the two critical damping points further apart, thus making it more difficult to find a convenient operating

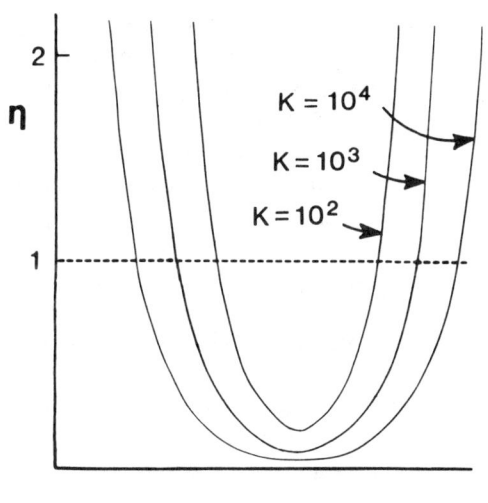

Fig. 3. Example of specific damping ratios (η) for different values of voltage-clamp control-amplifier gain K. Note that increasing K moves two values of critical damping ($\eta = 1$) further apart.

range. Therefore to achieve fast critical damping the experimenter must often employ the minimum gain consistent with an allowable static error.

If the system had a fixed resonance frequency, the response would be slower as the system became overdamped, and a damping ratio near critical damping would be the best compromise between sluggishness and oscillation. However, for the voltage-clamp system, ω_0 and η both depend on τ_i in such a way that ω_0 decreases as η increases; thus the system does not necessarily become slower as it becomes overdamped. We can show by expanding the response function for $\eta > 1$ that the actual rise time remains almost constant between the point of critical damping and infinite bandwidth. Substituting Equation 19 into Equation 18 and expanding for $\eta \gg 1$, the response becomes

$$V_m/V_c' \cong 1 - (1/2\ \eta^2 + 1) \exp(-Kt/2\tau_e) \qquad (23)$$
$$+ [1/2\ \eta^2 \exp(-Kt/2\tau_e)(2\eta^2 - 1/2)]$$

so that the rise time varies only slightly from critical damping to infinite bandwidth. Equation 23 closely approaches the infinite-bandwidth solution of Equation 11; thus the improvement in response speed above the point of fast critical damping is at most a factor of two (comparing Eq. 22 with Eq. 23). Furthermore there is no special minimum rise time for τ_i between $\tau_e/4K_0$ and 0 so that anywhere beyond the condition of fast critical damping the system will be fast and increasingly overdamped. As can be seen from Equation 23, these two effects tend to cancel.

The problem in clamp design for small cells is thus either to have a bandwidth that exceeds the fast critical damping criterion or to alter the system so that operation below the slow condition is not prohibitively sluggish. The goal is to achieve a sufficient separation between τ_i and τ_L to prevent oscillation when the system is operated at high gain. Exceeding the fast condition implies a rather small value for τ_i, which often cannot be physically realized. In the next section we discuss prototypical methods for changing the η-versus-τ_i curve to make the fast condition realizable. Consider as an example a spherical neuron 20 μm in diameter clamped with a 20-MΩ electrode resistance. For this case, τ_e would be 240 μs and the value of τ_i needed to reach critical damping is ~0.24 μs for $K_0 = 10^3$, requiring an amplifier bandwidth of 2.6 MHz or a gain-bandwidth product of 2.6 GHz, which may not be attainable with existing operational amplifiers. If the system, operated at full bandwidth, cannot reach the fast critical damping condition, the operating point will end up in the underdamped region and the best stable operation can only be obtained at or below the slow condition of critical damping. Because this may be prohibitively slow, some other "trick" must be used to improve the response time. For the moment, however, let us dwell on the limit imposed by Equation 22. One might ask whether it is possible to artificially increase τ_e so that a system of realizable τ_i can be made to reach the condition of fast critical damping. Examination of Equation 18 shows how $\eta(\tau_i)$ varies as R_e, and thus $\tau_e = R_e C_m$, is increased. In Figure 4 we see that increase in R_e shifts the $\eta(\tau_i)$ curve to the right (curve A to curve B), making the condition of fast critical damping more accessible. In the test experiment shown in Figure 5, τ_e was increased by inserting a high resistance (R_e) in series with the current output. However, as R_e is made

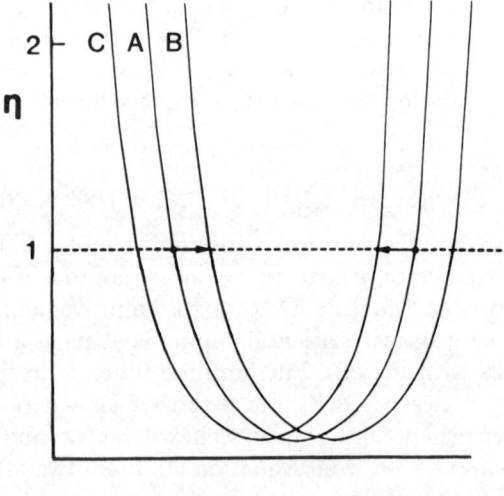

Fig. 4. Effect of changing effective electrode resistance (R_e in Fig. 1) on damping ratio (η). *Curve A* shows unaltered system with amplifier gain of 10^3 and $R_e = 10^7$ Ω. *Curve B* shows effect of increasing R_e by a factor of 10 as described in text. *Curve C* shows effect of current feedback cancellation of 90% of R_e. In B, fast critical damping point is moved into more physically realizable range of τ_f (*left arrow*). In C, slow critical damping is moved into range of faster overall response time (*right arrow*) but on slow limb of curve.

Fig. 5. Demonstration of increasing response time from critical damping (A) to slightly underdamped response (B) by increasing R_e (Fig. 4, *curve A* to *curve B*) and increasing system gain (Fig. 3, *curve* $K = 10^3$ to *curve* $K = 10^4$).

large the combination of control amplifier plus large series resistance becomes effectively a controlled current source (i.e., device that delivers current dependent only on error signal and independent of electrode resistance). Clamps of this sort have been used with linear voltage-regulated current sources [(15); see the chapter by Finkel and Redman]. The output of the current controller can be represented by

$$I = K_g(s)(V_c - V_m) \qquad K_g(s) = K_g(1 + s\tau_i)^{-1} \qquad (24)$$

where the transconductance (K_g) is assumed to have a finite bandwidth characterized by τ_i and I is current. The transfer function for the arrangement is

$$T(s) = K_g R_m / [K_g R_m + (1 + 1\ s\tau_m)(1 + s\tau_i) \qquad (25)$$

which again can be characterized in terms of ω_0 and η, where

$$\omega_0 = (1 + K_g R_m)(\tau_i \tau_m)^{1/2}$$

and

$$\eta = (\tau_i^{-1} + \tau_m^{-1})/(2\omega_0) \qquad (26)$$

When we write the condition of critical damping, $\eta = 1$, we now find for the fast condition

$$\tau_i = \tau_m/4(1 + R_m K_g) \qquad (27)$$

For a value $K_g = 10^2 R_m^{-1}$, which is easily realized, this system gives a rise time of 40 μs for $\tau_i = 5$ μs. Comparison of Equations 13 and 25 shows that the original voltage amplifier and high resistance is just a special case of Equation 25 with $K_g = K_0/R_e$.

Two–time-constant control amplifier

From the foregoing discussion, the description of the clamp as a second-order system appears to impose rather severe restrictions on clamp design because the two points of critical damping occur at such widely separate times, one too fast to be reached and the other too slow to be useful. In practice there is no need to have a control-amplifier response with such a simple frequency dependence as that of Equation 15. Smith et al. (16) show how a clamp can be designed with a two–time-constant control amplifier, so that

$$K(s) = K_0/(1 + s\tau_i) + K_1/(1 + s\tau_H)$$

where K_1 and τ_H are the added gain and time constant, respectively. To show how such a device works we can approximate by letting $K_1 \ll K_0$ and $\tau_H \cong 0$. For this case the transfer function of Equation 17 is altered in such a way that the system remains second order but the parameters are altered. In particular, the damping ratio becomes

$$\eta' = \eta + (K_1/K_0)(\tau_i/16\tau_f)$$

where τ_f is the original value, $\tau_e/4K_0$ (Eq. 21). This altered curve is plotted in Figure 6 for the values $K_0 = 10^3$ and $K_1 = 10$. From the figure it can be seen that the slow critical damping condition speeds up by an order of magnitude whereas the fast condition is hardly changed. The practical design of this type of clamp is discussed further in **Voltage Clamping Tissue-Cultured Neurons** (p. 245).

Improving clamp response with current feedback

An alternative method for compensating the IR drop across the current-delivering electrode is the use of current feedback (9). This is analogous to the series-resistance compensation devised by Hodgkin et al. (7) to make up for the IR drop across the Schwann cell layer of the giant axon; it is also analogous to the current feedback that is discussed for the whole-cell patch clamp (see **Whole-Cell Patch Clamp**, p. 250). The compensations differ, however, in the placement of the resistance that is compensated. In the series-resistance compensation employed for giant axons, the series resistance resides in the periaxonal tissue layer between the voltage-sensing electrode and the axolemma membrane. This series resistance is seen as part of the membrane and contributes

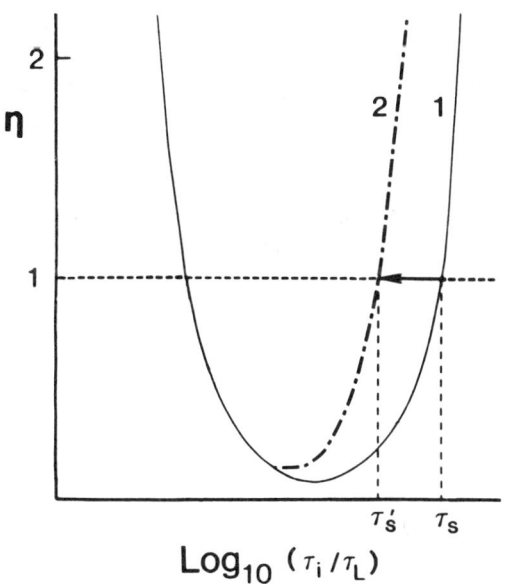

Fig. 6. Alteration of η curve by addition of wide-bandwidth low-gain response to voltage-clamp control amplifier. *Curve 1* shows control system with response time τ_s. *Curve 2* shows effect of adding new response. Result is increase in system response time (*arrow* to τ_s').

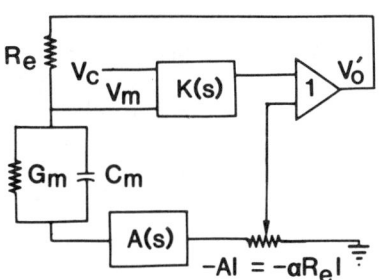

Fig. 7. Simplified schematic diagram of current feedback compensation by applying portion of membrane current $(-\alpha RI)$ to final stage of voltage-clamp amplifier system to produce altered clamp output V_0'. $A(s)$ represents frequency dependence of current-monitor amplifier. This electronic reduction of effective R_e mainly has effect of shifting η curve to *left* and increasing system speed (Fig. 4, *curve A* to *curve B*).

an IR drop that is outside the voltage-control loop. In the Katz and Schwartz (9) scheme, current feedback is used to overcome the IR drop across the current-delivering microelectrode, which is inside the voltage-control loop, as an adjunct to the voltage clamp, which does the same job but not speedily enough.

Katz and Schwartz (9) analyzed the effect of their current feedback for an approximate arrangement in which the feedback amplifier was infinitely wide band. Thus the predicted response is first order and shows a large decrease in rise time with no underdamping, as is characteristic of first-order approximate systems. In reality the very large improvement only comes when the rise-time parameter is nearly zero; further compensation would cause the rate constant to go negative so that operation near this limit would be unstable.

The effect of current feedback is shown in curve C of Figure 4. Figure 7, with $A(s)$ (response function of current-feedback amplifier) infinitely wide band, leads to the same transfer function as Equation 16 but with R_e replaced by a smaller value R'_e, which is adjustable. Figure 4 shows the shift for $R'_e = 0.1 R_e$. As can be seen in the figure the effect of current feedback is to shift the curve to the left, bringing the slow critical damping point into a more practical fast response.

When the bandwidth of the current feedback amplifier is taken into account, we add current feedback to the system of Figure 1, giving us the configuration of Figure 7. Letting τ'_I be the response time of the current amplifier, the transfer function is

$$T_I(s) = \frac{K(1 + s\tau_i)}{(K + 1)(1 + s\tau'_I) + R(1 + s\tau'_I) - (\alpha RG + s\tau_i)} \quad (28)$$

where αR is the adjustable current signal fed back to the control amplifier. If the experimenter adjusts $\alpha R = R_e$ to obtain perfect balance, the transfer function becomes

$$T_I(s) = \frac{K(1 + s\tau'_I)}{(K + 1)(1 + s\tau'_I) + (R_e/R_m)s\tau'_I(1 + s\tau_m)(1 + s\tau_I)} \quad (29)$$

The case discussed by Katz and Schwartz (9) corresponds to $\tau'_I = 0$, so that $T_I(s) = K/(K + 1)$. One can consider a situation in which the system has been tuned to slow critical damping, i.e., $\tau_i = 4K\tau_e$. By substituting this value into Equation 28, one can study the stability of the resultant third-order system as a function of τ'_I. The denominator of the transfer function can be examined according to the Routh criterion (17). This denominator is

$$B(s) = [(R_e/R_m)\tau'_I\tau_m 4K\tau_e s^3 + (R_e\tau'_I)R_m + \tau_m R_e \tau'_I/R_m^2]s \\ + [(R_e/R_m)\tau'_I + (K + 1)\tau'_I]s + (K + 1) \quad (30)$$

which can be simplified for large K so that we can study the cubic form

$$s^3 + \tau_e(R_e/R_m)s^2 + (1/4\ \tau_e^2)s + (1/4\ \tau_e^2 \tau'_I) \quad (31)$$

which will give a condition analogous to critical damping only if all three roots have positive real parts. The Routh criterion requires that certain coefficients not change sign. The first two coefficients are $4K\tau_e$ and 1, which are intrinsically positive; however, the third coefficient of the Routh array is $4K\tau_e(1 - 4K\tau_e/\tau'_I)$, which will be negative for any reasonable value of K. The fourth coefficient, $K/\tau_i\tau'_I$, is positive so the cubic form has at least two unstable roots, unless the third term is positive, which occurs when $4K\tau_e/\tau'_I < 1$ or for the response times longer than $\tau'_I = 4K\tau_e$, which is just the condition of slow critical damping. Thus the analysis predicts no obvious improvement. Probably some improvement is obtained when τ_i is varied to give an underdamped response and τ'_I is then varied independently. We do not discuss this case, but empirically one usually finds small (factor of 2) improvements and not the enormous improvements predicted for the infinite-bandwidth case.

How to realize fast rise time

Note that τ_e and τ_m scale differently with cell size. For an isopotential cell, R_m is inversely proportional to surface area and C_m is directly proportional to area; thus τ_m should be independent of cell size. The main reasons that τ_m may vary for different cells are either that some cells have extensive dendritic networks or that R_m differs considerably for different cells at rest. The time constant τ_e, on the other hand, depends on cell size; it becomes slower for larger cells. Thus to realize the ideal rise time for a small cell (assuming $\tau_e \ll \tau_m$) we must have a system response time $\tau_i = \tau_L/4K$. This can be achieved by *1*) speeding up the clamp, *2*) somehow slowing the load, or *3*) modifying the clamp to have a different criterion for optimum rise time. One example of how a clamp is modified in practice is given in the next section.

Voltage Clamping Tissue-Cultured Neurons

Vertebrate central nervous system cells grown in culture, which possess the electrophysiological characteristics of neurons found in vivo, have a number of properties that make them an attractive research preparation. These properties (e.g., that the cells can be directly visualized) make it practicable to undertake experiments that are difficult or impossible on neurons in situ. Such experiments include voltage clamping with two microelectrodes. For experiments that clamp only slow events, like those in which membrane responses to bath or iontophoretic neuroactive substances are investigated, the conventional approach described in the chapter by Finkel and Gage is quite adequate. However, when there is a need to clamp fast events, like spikes, the conventional approach usually will not work.

Critical damping requires a considerable difference between τ_i and τ_L. With most preparations, this means making τ_i as small as possible because τ_L is usually large (16). In tissue-cultured neurons, however, τ_L is quite small due to the small value of C_m. This leads to a situation whereby physical limitations in electronic devices make it impossible to achieve a sufficiently small τ_i to realize the necessary separation of time constants to obtain fast critical damping. Critical damping can be obtained by increasing τ_i by slowing down the electronics, but this defeats the goal of clamping fast conductance changes. In practice this was achieved by placing a capacitor (C_f) in parallel with the gain-control potentiometer (R_g) in the feedback loop of an operational amplifier (see

Fig. 8, *inset*). However, it was found that fast clamping could be achieved by adding an additional time constant in the clamping amplifier (Fig. 9). This additional time constant was obtained by adding a potentiometer (R_f) in series with C_f. How this result comes about can be understood from an analysis of the Bode plots in Figures 8 and 10, where the logarithm of the clamp gain and phase are plotted against the logarithm of frequency. Adjustment of the mixed curve is illustrated in Figure 10. When $R_f = 0$, the result is a pure slow system shown as the left-most curve. As R_f is increased, the Bode plot changes progressively through curves 1 and 2 and R_f becomes very large; curve 3 or a pure fast system results because the filtering effects of C_f have been removed. Curve M_g in Figure 8 represents a typical setting for clamping cultured spinal neurons. The mixed system maintains the large difference between τ_L and τ_i; it also maintains a large overall value for amplification. However, it provides a low-gain high-frequency boost to the system and retards the phase shift (M_p). This makes it possible to clamp fast events reasonably well without oscillations (Fig. 9).

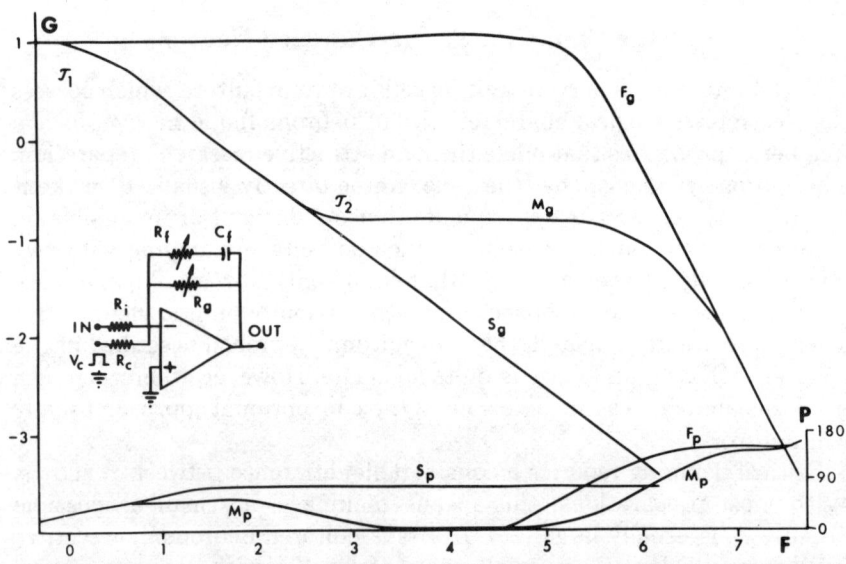

Fig. 8. Bode plots of gain-frequency control amplifier in two-microelectrode voltage-clamp amplifier. *Inset*: circuit design of gain-frequency control amplifier. Pulse into negative input is command pulse; R_i, input resistor that receives V_m from headstage; R_c, input to which command signal V_c is applied; R_g, gain control potentiometer; R_f, frequency response potentiometer; C_f, frequency response capacitor; F_g and F_p, gain and phase curves, respectively, for fast clamp system; S_g and S_p, gain and phase curves, respectively, for slow clamp system; M_g and M_p, gain and phase curves, respectively, for mixed clamp system. Graphs are log gain (G) versus log frequency (F), with numbers indicating exponents of 10, and phase shift (P) in degrees versus log F.

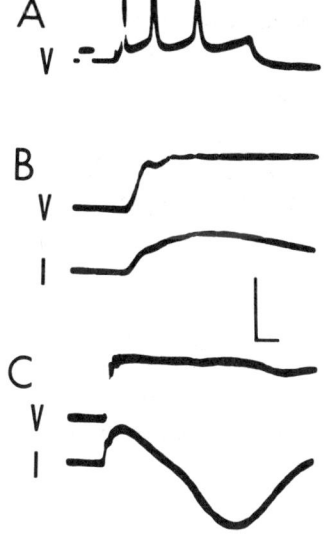

Fig. 9. Intracellular recordings from tissue-cultured mouse spinal cord cell. V, membrane voltage; I, membrane current. A: multiple action potentials evoked by 100-ms depolarizing constant-current pulse. B and C: two-microelectrode voltage-clamp recordings from cell illustrated in A. Membrane stepped from holding potential of −60 to −37 mV. Calibrations: *vertical bar* is 50 mV in A, 25 mV in *upper records* and 0.125 µA in *lower records* of B and C; *horizontal bar* is 10 ms in A, 20 µs in B, and 100 µs in C.

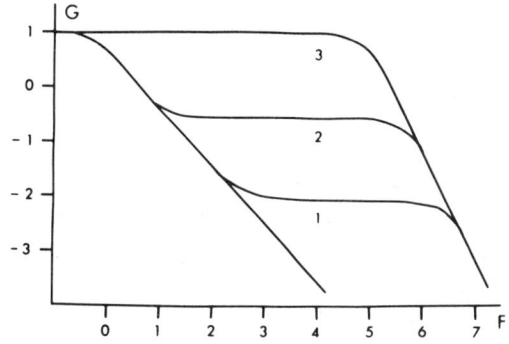

Fig. 10. Effects of changing R_f in Fig. 8.

Space-Clamp Considerations

How small is small? In the design of voltage-clamp experiments a great amount of ingenuity and analysis has been devoted to overcoming the tyranny of cell geometry. If the observed ionic currents are to represent the dynamics of membrane processes, there has to be an assurance that the membrane potential is reasonably uniform over the area from which current is collected. Most methods discussed in this book presuppose either isolated patches of membrane or whole cells small enough for the membrane to have an approximately isopotential surface. For the single- and two-microelectrode clamps, no "tricks" are discussed for obtaining space-clamp conditions; the experimenter must know when the cell geometry allows valid space clamp. In this section we review order-of-magnitude estimates of maximum cell size for four cell geometries: short cylindrical cables, spherical cells, tissue sheets (or pancake-shaped cells), and dendritic networks (see the chapter by Rall).

Short cylindrical cell

For a short cylindrical cell with sealed ends, such as a myotube, the potential response in space and time to a current step at $x = 0$ is

$$V(x,t) = V'_0 \frac{\cosh(l-x)/L}{\cosh(l/L)} + \text{(time varying terms)} \qquad (32)$$

where L is the length constant and l is the length of the cell (1). The response becomes slower and falls off with distance (x) from the current-delivering electrode. If we consider only the steady-state behavior given by the first term on the right, we can ask how long the cell can be to have $V(x)$ above a prescribed value, say 90% of V'_0. Setting $V(l) = 0.9K_0$, we obtain

$$l(90\%) = 0.47L \qquad (33)$$

For a one-dimensional cable

$$L = [r_m/(r_e + r_i)]^{1/2} = (d/4R_m/R_i)^{1/2} \qquad (34a)$$

where d is the diameter of the cell, r_m is membrane resistance per unit length, r_e and r_i are external medium and cytoplasmic longitudinal resistances per unit length, respectively, and R_m and R_i denote membrane resistance of a unit area and cytoplasmic resistivity, respectively. The maximum permissable length scales as $d^{1/2}$ provided that $d \ll l$ and that the one-dimensional cable approximation is valid. For a typical myotube at rest, $d = 20$ μm, $R_m = 10^4$ Ωcm, and $R_i = 100$ Ωcm; therefore $d_{\max} = 1,500$ μm. Thus for an experiment in which acetylcholine-induced current fluctuations are measured, a myotube would be under reasonable spatial control. For a small axon, excitation might decrease R_m by a factor of 100 or more, and space clamp would become problematic. For an electrically excitable cell, the cable equation should be replaced by a nonlinear equation; however, because many excitable conductances increase with voltage according to a sigmoid relation, a lower limit to the length constant can still be obtained from Equation 34 if L is replaced by an "active" length constant

$$L_{\text{act}} = (L^{-2} + r_i g_e)^{1/2} \qquad (34b)$$

where g_e is the maximum conductance per unit length contributed by the nonlinear conductance. Equation 33 with L_{act} substituted from Equation

34b suggests a length of the order of only 150 μm for an active fiber. This is only seven times the diameter; thus the fiber cannot really be considered to be a one-dimensional cable. To estimate the error produced by departure from space clamp, one can calculate the current response to a step at $x = 0$ (and therefore voltage of the form given by Equation 32) and integrate over distance comparing the current from nonclamped regions with the current from regions near the electrode.

Spherical soma

Hellerstein (6) and Eisenberg and Engel (2) have shown that a spherical soma is approximately isopotential for cells of any reasonable size. In Hellerstein's formulation, for example, the membrane potential in response to current delivered to a point on the surface is expanded in a series of Legendre polynomials. The ratio of the first angular-dependent term (V_1) to the isotropic term (V_0) is given by

$$V_1/V_0 = 3/2 \; d(R_e + 2R_i)/R_m \tag{35}$$

For $R_m = 4 \times 10^3 \; \Omega\text{cm}^2$ and $R_i = R_e = 50 \; \Omega\text{cm}$, as expected for a molluscan neuron at rest, the cell diameter for a 10% error in potential is 1.8 cm, far larger than the diameter of any cell of interest. Thus the real spatial problem in clamping small cells without processes does not arise from space-clamp difficulties but from other factors associated with size, such as the proximity of two microelectrodes or the capacitance-to-ground shunts for high-impedance microelectrodes.

Pancake-shaped cell or syncitial sheet

For a configuration of two parallel-plane sheets of membrane, Jack et al. (8) have shown that the potential falls off roughly as $\exp(-Lr)/Lr$, where r is the radial distance and L is the space constant given by

$$L = (R_m b/2R_i)^{1/2}$$

and where b is the distance between the two membranes. If the potential is to be homogeneous within 10%, the maximum radius would be $r = 0.6L$. For $b = 10 \; \mu\text{m}$, as might occur in a sheet of epithelial cells and for R_m and R_i as before, $L = 2$ mm. For 10% accuracy the maximum cell radius is 1,200 μm. Thus, just as for a spherical soma, a disklike soma of any practical size is likely to be space clamped even if the soma is highly excitable.

Cells with dendritic networks

When an isopotential soma has a passive dendritic network, the soma is still isopotential with somatically placed electrodes, but a large amount of current can be drained into the dendrites. This decreases the ratio of membrane resistance to electrode resistance, which is an important factor in single-electrode voltage clamping (see the next section). The large drain to the dendrites also means that the voltage clamp must deliver more current in order to control the membrane potential. This situation has been discussed thoroughly by Rall (14). As an example, consider a spherical soma with a few noticeable long dendrites. The conductance of the soma (G_s) is given by

$$G_s = \pi d_s^2 / R_m \qquad (36)$$

and the input conductance (G_d) of a cablelike dendrite is

$$G_d = (R_m^{-1} R_i^{-1} 4\pi d^3)^{1/2} \tanh(l/L) \qquad (37)$$

Because L for a 2-μm-diam dendrite is 700 μm, most dendrites would be much shorter than a length constant. A 100-μm-long dendrite of 2-μm diameter, for example, would contribute 0.6 nS of conductance or the equivalent of a 14-μm-diam spherical soma. As has been shown by Rall (14), dendritic networks may have 10–40 times the conductance of the neuron soma, making the current drain for a moderate voltage change prohibitive. Seldom can the steady-state admittance of a dendritic network be ignored. On the other hand, for transient responses evoked at the soma (e.g., with iontophoretically applied neuroactive compounds) when the membrane potential is controlled with intrasomatic electrodes, the presence of the dendrites is less of a problem.

Whole-Cell Patch Clamp

In addition to making single-channel observations possible, the gigohm seal of patch electrodes to cell membranes allows a low-impedance access to the cell interior (3, 13). Hamill et al. (5) illustrate how the membrane patch isolated by a gigohm seal can be ruptured so that the patch electrode becomes an internal electrode. This is a particularly convenient variant on the suction-pipette method of voltage clamping introduced by Lee et al. [(11); see also the chapter by Brown et al.] and Kostyuk and Krishtal (10) for large cells. If a step of potential is delivered to a membrane without feedback from a sensor of the true membrane potential, the membrane is at best only approximately voltage clamped. The pipette resistance is in series with the membrane impedance, and there is a voltage drop across the pipette that is directly proportional to the current. In this case the potential across the membrane will not be a step function because the membrane capacitance charges with a response time constant τ_e. The advantage of the "ruptured-patch" method is that it allows a reduced R_e. The advantage of small cells is that they have a small C_m. The electronics for this technique can have one of two configurations: *1*) the *IV* converter of the conventional patch clamp (see the chapter by Auerbach and Sachs) can be employed as the headstage of the system (Fig. 11) or 2) an electrometer can be used (Fig. 12). In both systems it is essential that the membrane resistance be very much larger than the electrode resistance; the latter must be known, must be properly compensated, and must remain constant.

The advantage of the Dagan-type system (Fig. 11) is that it exploits the low-noise current monitor. Its disadvantage is that the membrane voltage is never actually measured and compared with the command voltage to provide a difference feedback signal. In that sense, it is different from a conventional voltage-clamp system. The advantage of the WPI-type system (Fig. 12) is that it actually performs this latter function. Its disadvantage is that it may not have a low-noise current monitor. The potential advantage of either instrument is the use of a single low-resistance microelectrode. The disadvantages also relate to the microelectrodes. First, it is not certain when the current feedback to compensate for the *IR* drop across the microelectrode is set correctly. Second, if the value of the microelectrode resistance changes (usually from passing large currents or from closing of electrode tip), the membrane potential or potential change will not be the nominal value but some unknown value. The Dagan-type patch-clamp amplifier uses the same operational amplifier for measuring membrane current and for setting the potential of the inside of the electrode equal to a control signal. When operating in the whole-cell mode, these devices use a lower value of feedback resistance than when operating as single-channel detectors.

Let us first consider the limitations on a single-electrode clamp operating without feedback. The commercial patch clamps have current feedback compensation for the voltage drop across the electrode and we discuss the limitations of this type of control at the end of this section.

The membrane potential V_m, which is not directly accessible to measurement, responds to a step command V_c as

$$\frac{dV_m}{dt} + \frac{(1 + G_m R_e) V_m}{R_e C_m} = \frac{V_c}{R_e C_m} \tag{38}$$

which when integrated becomes

$$V_m = [V_c R_m/(R_m + R_e)][1 - \exp(-t/\tau_E)] \tag{39}$$

where $\tau_E = R_e C_m/(1 + G_m R_e)$. The expected voltage error and rise time depend on electrode and cell resistances and thus on cell size. For a fixed-size patch electrode we can see how the rise time and voltage error scale with cell size. Recall that $C_m = C'_m A$ and $R_m = R'_m/A$, where the primed quantities are specific values and A (membrane area) is πd^2 for a spherical cell. Substituting into Equation 39, the rise time becomes $R_e C'_m A/[1 +$

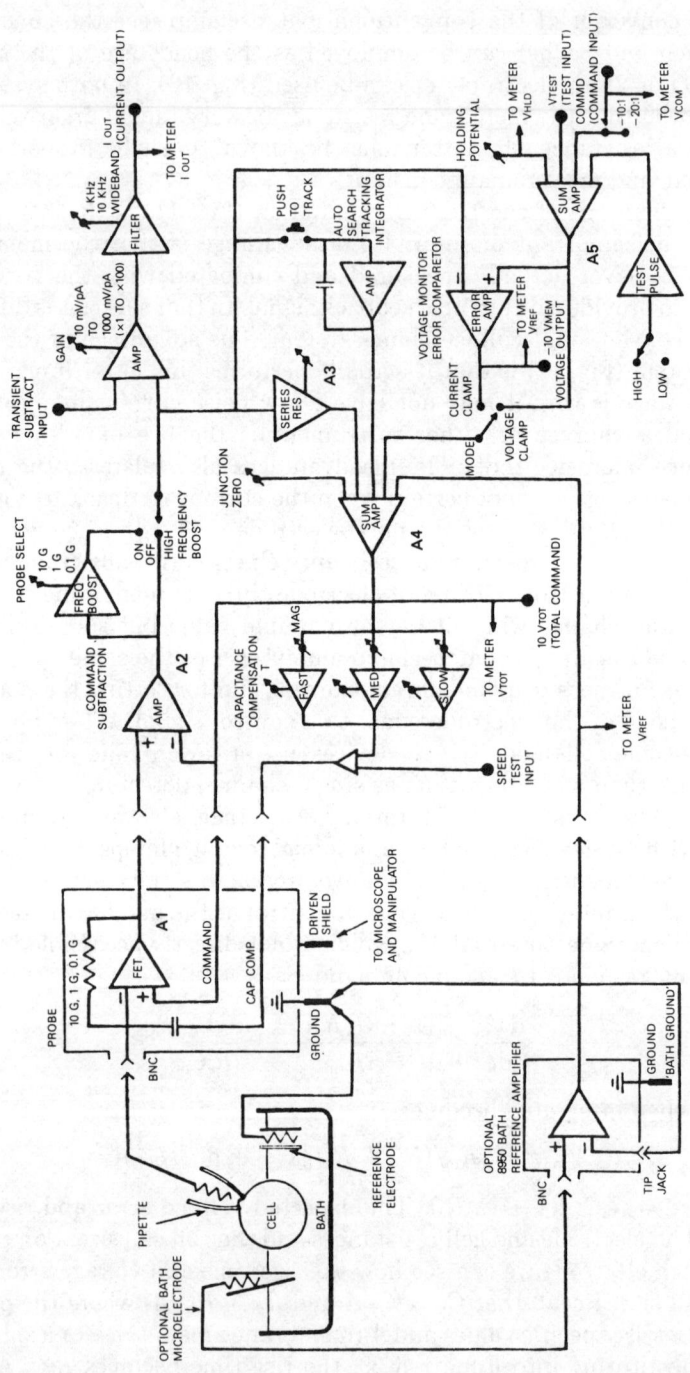

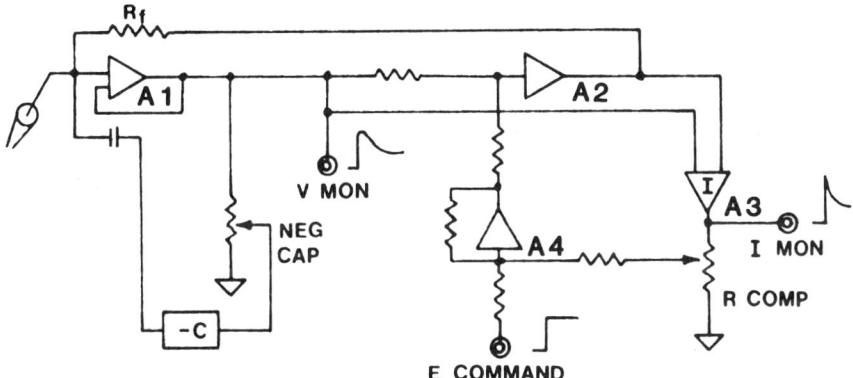

Fig. 12. Schematic diagram of WPI-type whole-cell patch clamp. Input stage in this system (A1) is electrometer and records signal that is directly related to membrane potential (V MON). Note that this signal contains membrane potential plus IR drop across microelectrode. Membrane potential could be obtained by subtracting this IR drop (R COM) from V MON. In addition, it has a means of compensating input capacitance of A1 (NEG CAP and $-$C). Membrane current is monitored as IR drop across feedback resistor R_f with A3 (I MON). As in Dagan-type system (Fig. 11), it adds signal proportional to IR drop across microelectrode (R COMP) to command signal (E COMMAND) to summing amplifier to try to make membrane potential equal to command potential. (Reproduced with permission of World Precision Instruments, Inc., New Haven, CT.)

$R_e(R'_m)^{-1}A]$, which approaches τ_e as $A \to 0$ and $\tau_m \to \infty$. To keep the rise time <50 μs, for example, we would be limited to cells with diameters <18 μm (assuming $R_e = 5$ MΩ). Similarly, to keep the voltage error $(1 + R_m/R_e)^{-1}$ under 0.05, we need a maximum cell diameter of 60 μm. Thus a patch clamp with no provisions for correcting the access-resistance error will only work on very small cells and/or with very large electrodes. This consideration is probably a bit optimistic since we have not discussed electrically excitable elements in the membrane, which place a premium on fast response time and accurate voltage, nor taken into account cells with dendrites for which the effective R_m is even smaller.

Fig. 11. Schematic diagram of Dagan-type whole-cell patch clamp. Note that input stage (A1) does not monitor membrane potential but is current-to-voltage converter of membrane current. Its output also contains any command (COMMAND) signal, but this is subtracted out by amplifier A2. Signal proportional to membrane current (A3 output) and intended to be equal to IR drop across patch electrode (PIPETTE) is added to final summing command amplifier (A4), which also receives membrane voltage command signal from A5. Thus command signal delivered to headstage (COMMAND of A1) is sum of electrode IR drop and intended membrane potential change. Note also that there is no direct measurement of resting or changing membrane potential in this system. It assumes that initial membrane voltage command signal (COMMAND of A5) is membrane potential change. There is no way to determine resting potential with this system. (Reproduced with permission of Dagan Corp., Minneapolis, MN.)

Paradigm for use of current feedback in single-electrode arrangement

The situation discussed so far depends on minimizing the access-resistance error, but there has been no discussion about compensation of this error by electronic feedback. Such current feedback loops are built into the single-electrode voltage clamps. Figure 12 shows a schematic diagram of one of the commercial patch-clamp amplifiers as used for whole-cell clamping. Amplifier A2 compares the electrode potential (V_e) with the input command signal (V_c) and thus serves the function of the control amplifier in the usual voltage clamp (although controlling a different voltage, $V_m + V_e$ instead of V_m). Amplifier A3 gives an output proportional to $V_e - V_0$ and therefore proportional to the current. Because the electrode potential differs from the true membrane potential by the amount $I_m R_e$, a fraction of the output proportional to $I_m R_e$ can be added to V_c to make the control of the membrane potential nearer to ideal. This is a positive current-feedback loop, and in practice it does improve matters, but not perfectly (nothing is perfect).

The circuit shown in Figure 12 can be studied as a paradigm for the use of current-feedback to cancel the errors produced by the $I_m R_e$ drop. Note also that the correction, as pictured here, is made at the outset of an experiment and that there is no provision for subsequent electrode drift or polarization, which requires an intermittent control in which the error can be sampled continuously (see the chapter by Finkel and Redman). In Figure 12, we assign a response time (τ_i) to the summing amplifier. If the amplifier were infinitely wide band, the correction for series resistance could be made virtually perfect (although the resulting network would verge on instability and some limitations would be imposed by the necessity to reduce the correction). With a finite bandwidth the output of the summing amplifier can be written as

$$V_0 = (V_c + BI)/(1 + s\tau_i) \tag{40}$$

where B is an arbitrary resistive setting that is varied by the experimenter. Because $V_m = V_0 - I_m R_s$, we can substitute for V_0. Then for the resting membrane, letting

$$I_m = Y_m V_m = G_m(1 + s\tau_m)V_m \tag{41}$$

the transfer function for this network can be written as

$$V_m/V_c = (1 + s\tau_i)/(G_m R_e \tau_i \tau_m s^2 \\ + \{(1 + G_m R_e)\tau_i + [C_m(R_e - B)\tau_m]s + G_m(R_e - B)\}) \tag{42}$$

We can solve this for the condition of critical damping, or for simplicity we can first set $B = R_e$ as one would do in adjusting the real clamp and then solve for the vlaue of T, which gives critical damping. After some algebra, this leads to

$$\tau_i = 4\tau_m G_m R_e/(1 + G_m R_e)^2 = 4\tau_e/(1 + G_m R_e)^2 \qquad (43)$$

For values of τ_i less than the value in Equation 43, one achieves the best possible cancellation, provided the clamp can deliver the required current. In practice, as the value of B is varied the second term of the denominator of Equation 42 can go negative and the system may oscillate. This calculation illustrates the kind of finite bandwidth effects that set the limit on this type of current feedback. The calculation can be considered to be a simplified representation of the workings of a variety of similar schemes.

Conclusion

In this chapter we attempted to cover some of the topics not covered in the other chapters of this book and areas of voltage clamping that have arisen since this book and the workshop from which it derived were planned. These include the subject of clamping small cells and particularly the technique called whole-cell patch clamping. This name, while somewhat misleading, involves voltage clamping small cells with a relatively large low-resistance single electrode. It promises to be a major new tool in the armamentarium of electrophysiologists because it allows the investigation of the electrical properties of a wide variety of cells heretofore unapproachable by electrical techniques.

Also we introduced a somewhat novel approach to analyzing voltage-clamp systems in terms of clamp stability, static errors, dynamics, and speed of response. We hope this provides voltage clampers with a rational, mathematical, and graphic means of assaying the clamping of any particular cell in an optimum way. We also hope some of the suggested modifications or "tricks" can be applied to achieve the desired results in a logical manner. In addition, we gave some approximations to the adequacy of space clamping for cells of different geometries.

REFERENCES

1. Bezanilla, F., J. Vergara, and R. E. Taylor. Voltage clamping of excitable membranes. In: *Methods of Experimental Physics, Biophysics 20*, edited by G. Ehrenstein and H. Lecar. New York: Academic, 1982, p. 445–511.
2. Eisenberg, R. S., and E. Engel. The spatial variation of membrane potential near a small source of current in a spherical cell. *J. Gen. Physiol.* 55: 736–757, 1970.
3. Fenwick, E. M., A. Marty, and E. Neher. A patch-clamp study of bovine chromaffin cells and of their sensitivity to acetylcholine. *J. Physiol. London* 331: 577–597, 1982.
4. Guttman, R., L. Feldman, and H. Lecar. Squid axon membrane response to white noise simulation. *Biophys. J.* 14: 941–955, 1974.
5. Hamill, O. P., A. Marty, E. Neher, B. Sakmann, and F. J. Sigworth. Improved patch-

clamp techniques for high-resolution current recording from cells and cell-free membrane patches. *Pfluegers Arch.* 391: 85–100, 1981.
6. Hellerstein, D. Passive membrane potentials. A generalization of the theory of electrotonus. *Biophys. J.* 8: 358–379, 1968.
7. Hodgkin, A. L., A. F. Huxley, and B. Katz. Measurement of current-voltage relation in the membrane of the giant axon of *Loligo. J. Physiol. London* 116: 424–448, 1952.
8. Jack, J. J. B., D. Noble, and R. W. Tsien. *Electric Current Flow in Excitable Cells.* Oxford, UK: Clarendon, 1975.
9. Katz, G. M., and T. L. Schwartz. Temporal control of voltage-clamped membranes: an examination of principles. *J. Membr. Biol.* 17: 275–291, 1974.
10. Kostyuk, P. G., and O. A. Krishtal. Separation of sodium and calcium currents in the somatic membrane of mollusc neurons *J. Physiol. London* 270: 545–568, 1977.
11. Lee, K. S., N. Akaike, and A. M. Brown. Properties of internally perfused, voltage-clamped, isolated nerve cell bodies. *J. Gen. Physiol.* 71: 489–508, 1978.
12. Levis, R. Patch and Axial Wire Voltage-Clamp Techniques and Impedance Measurements of Cardiac Purkinje Fibers. Ann Arbor, MI: University Microfilm, 1981. PhD thesis.
13. Marty, A., and E. Neher. Tight-seal whole-cell recording. In: *Single-Channel Recording,* edited by B. Sakmann and E. Neher. New York: Plenum, 1983, p. 107–122.
14. Rall, W. Branching dendritic trees and motorneuron membrane resistivity. *Exp. Neurol.* 1: 491–527, 1959.
15. Sachs, F., and H. Lecar. Acetylcholine-induced current fluctuations in tissue-cultured muscle cells under voltage clamp. *Biophys. J.* 17: 129–143, 1977.
16. Smith, T. G., Jr., J. L. Barker, B. M. Smith, and T. R. Colburn. Voltage clamping with microelectrodes. *J. Neurosci. Methods* 3: 105–128, 1980.
17. Trimmer, J. D. *Response of Physical Systems.* New York: Wiley, 1950.

Index

Axons
  nonmyelinated, sucrose-gap voltage clamp for, 218–221
    advantages, 220–221
    problems, 221

Dendrites: *see* Neurons

Electrodes
  see also Microelectrode shielding
  ion-selective and iontophoretic electrodes in voltage clamping, shielding for, 31–32

Feedback
  current feedback for improving speed of response of two-electrode voltage clamp, 242–245
  current feedback in single-electrode arrangement for whole-cell patch clamping of small cells, 254–255
  linear feedback network, two-electrode voltage clamp as, 233–234

Ganglia
  circumesophageal ganglia of *Helix aspersa*, study of neurons by suction-pipette method of voltage clamp and internal perfusion, 163–166

Intestinal mucosa
  arteriolar smooth muscle of, voltage clamping with high-resistance microelectrode in guinea pigs, 118–119

Ion channels
  single-channel data acquired by high-resolution patch clamp, analysis of, 137–148
    data acquisition, 139–140
    estimating event amplitudes, 145–147
    estimating event durations, 142–143
    estimating number of active channels in patch, 147–148
    fitting histograms, 143–145
    low-pass filtering and event detection, 140–142
    theoretical framework, 137–139
  single-channel data, recording with high-resolution patch clamp, 122–129
    cell treatments, 122–123
    characteristics of patch-clamp data, 127–129
    pipette construction, 123–125
    seal formation and patch configuration, 125–127

Iontophoresis
  iontophoretic electrodes in microelectrode voltage clamping, shielding for, 31–32

Membrane channels: *see* Ion channels

Membrane potentials
  control by microelectrode and sucrose-gap voltage-clamp techniques, comparison of, 221–226
    criterion for voltage clamp, 225
    general concepts, 221–224
    membrane resistance-to-access resistance ratio, 224–225
    spatial voltage uniformity, 225–226
    speed of response, 225
  control by two-electrode voltage clamp, 47–94
    basic description, 54–56
    cell model, 53–54
    equivalent circuit, 59–60
    general considerations, 83–92
    ideal voltage clamp, 56–65
    membrane capacitance as sole frequency-dependent component, 56–65
    membrane current responses, 60–64
    membrane voltage responses to changes in command potential, 56–59
    real voltage clamps, 65–83
    terminology and methods, 48–54
  excitatory junction potentials in arteriolar smooth muscle in guinea pig intestinal mucosa, recording by voltage clamping with high-resistance microelectrode, 118–119
  measurement with single-electrode voltage clamp, 95–120
    principles of operation, 96–107
    steady-state ripple in membrane potential, 105–106
  membrane current measurement with microelectrode voltage clamping, 16–20
    series current measurement, 19–20
    virtual-ground current measurement, 16–18

258　Voltage Clamping

Microelectrode shielding
　automation of, 42–45
　　construction and operation, 42
　　controller operation, 42–45
　　detailed circuit description, 45
　　results, 45
　electrode holders, 40–41
　for ion-selective and iontophoretic
　　electrodes, 31–32
　for voltage recording, 26–28
　resistance requirements for good shields,
　　32–33
　shield construction, 34–38
　　mass-production shielding, 35–38
　　painted shields, 35
　　reusable shields, 35
　shield insulation requirements, 33–34
　shield insulation techniques, 38–40
　　glass insulation, 38–40
　　painted insulation, 40
　　thermosetting insulation, 38
　to decrease interelectrode coupling,
　　29–31
　　effects of coupling capacity, 29–31
　　effects of current-electrode capacity,
　　　31
　　effects of voltage-electrode capacity,
　　　31
Microelectrode voltage clamp: *see* Voltage-
　　clamp techniques
Motoneurons
　of spinal cord of cat, voltage clamping
　　with shielded single electrode,
　　116–118
Muscle, vascular smooth
　in guinea pig intestinal mucosa, voltage
　　clamping with high-resistance
　　microelectrode, 118–119

Neurons
　see also Axons; Motoneurons
　branched neurons studied by voltage
　　clamping with intracellular
　　microelectrodes, space-clamp
　　problems in, 191–215
　cable aspects of dendrites, 191–193
　current transfer from dendrites to
　　clamped soma, 203
　dendritic modulus and phase in AC
　　steady state, 199–202
　dendritic transfer in AC and Laplace
　　transform domains, 198–199
　relation between brief synaptic current
　　and current transient at voltage
　　clamp, 205–207
　relation to AC impedances, 202–203
　relevance to dendritic synapses,
　　193–194

　reversal potential for dendritic
　　synapses, 196
　steady-state voltage decrement with
　　distance, 194–196
　transfer in Laplace transform and
　　time domains, 203–205
　transient solution for voltage step at
　　soma, 196–198
　in tissue culture, voltage clamping of,
　　245–247
　of circumesophageal ganglia of *Helix
　　aspera*, study by suction-pipette
　　method of voltage clamp and
　　internal perfusion, 163–166

Patch-clamp techniques
　instrumentation, 130–137
　　changing patch potential, 135–136
　　frequency response, 134–135
　　noise sources, 130–134
　　recording potentials, 137
　　tracking (autozero) operation, 136–137
　single-channel data, analysis of, 137–148
　　data acquisition, 139–140
　　estimating event amplitudes, 145–147
　　estimating event durations, 142–143
　　estimating number of active channels
　　　in patch, 147–148
　　fitting histograms, 143–145
　　low-pass filtering and event detection,
　　　140–142
　　theoretical framework, 137–139
　single-channel data, recording of,
　　122–129
　　cell treatments, 122–123
　　characteristics of patch-clamp data,
　　　127–129
　　pipette construction, 123–125
　　seal formation and patch
　　　configuration, 125–127
　whole-cell patch clamp for small cells,
　　250–255
　paradigm for use of current feedback
　　in single-electrode arrangement,
　　254–255
Purkinje fibers
　microelectrode voltage-clamping studies
　　of, 171–189
　　accuracy and limitations, 174–175
　　clamp circuit, 184
　　clamp circuit modifications, 186–188
　　optimizing current microelectrodes,
　　　184–186
　　optimizing electrode spacings, 176–178
　　testing the clamp, 188
　　theoretical framework, 171–183
　　three-electrode technique, 174–175
　　time-dependent changes for linear and

nonlinear conductances, 178–182
two-electrode technique, 172–174, 184–189

Shielding of microelectrodes: *see* Microelectrode shielding
Single-electrode voltage clamp (SEVC): *see* Voltage clamping, single electrode
Spinal cord
  motoneurons of, voltage clamping with shielded single electrode in cat, 116–118
Sucrose-gap voltage clamp: *see* Voltage-clamp techniques
Suction-pipette method: *see* Voltage-clamp techniques

Tissue culture
  of neurons, voltage clamping of, 245–247
Two-electrode voltage clamping (TEVC): *see* Voltage clamping, two electrode

Voltage-clamp techniques
  see also Patch-clamp techniques; Voltage clamping, single electrode; Voltage clamping, two electrode
  for small cells, 231–256
    cells with dendritic networks, 249–250
    pancake-shaped cell or syncitial sheet, 249
    short cylindrical cell, 248–249
    space-clamp considerations, 247–250
    spherical soma, 249
  for tissue-cultured neurons, 245–247
  history and development of, 1–7
  in study of branched neurons, space-clamp problems in, 191–215
    cable aspects of dendrites, 191–193
    current transfer from dendrites to clamped soma, 203
    dendritic modulus and phase in AC steady state, 199–202
    dendritic transfer in AC and Laplace transform domains, 198–199
    relation between brief synaptic current and current transient at voltage clamp, 205–207
    relation to AC impedances, 202–203
    relevance to dendritic synapses, 193–194
    reversal potential for dendritic synapse, 196
    steady-state voltage decrement with distance, 194–196
    transfer in Laplace transform and time domains, 203–205
    transient solution for voltage step at soma, 196–198
  in study of cardiac Purkinje fiber, 171–189
    accuracy and limitations, 174–175
    experimental guidelines, 184–189
    theoretical framework, 171–183
    three-electrode technique, 174–175
    time-dependent changes for linear and nonlinear conductances, 178–182
    two-electrode technique, 172–174, 184–189
  microelectrode and sucrose-gap voltage-clamp techniques, comparison in compensation for series resistance, 226–230
    compensation for series resistance, 227–229
    effects and oscillations with compensation, 226–227
    gating currents, 229
    measurement of series resistance, 227
  microelectrode and sucrose-gap voltage-clamp techniques, comparison in control of membrane potential, 221–226
    criterion for voltage clamp, 225
    general concepts, 221–224
    membrane resistance-to-access resistance ratio, 224–225
    spatial voltage uniformity, 225–226
    speed of response, 225
  microelectrode and sucrose-gap voltage-clamp techniques, comparison of, 217–230
  sucrose gap for nonmyelinated axons, 218–221
    advantages, 220–221
    problems, 221
  suction-pipette method of voltage clamp and internal perfusion, 151–169
    circuit and cell characteristics, 160–163
    design considerations, 152–155
    practical tests of adequacy of exchange, 155–158
    preparation of medium-sized neurons, 163–166
    time resolution and spatial control of voltage clamp, 160
    voltage clamp, 158–159
  useful circuits for clamping with microelectrodes, 9–24
    adjustment of holding level before clamping, 23–24
    design of high-speed low-noise headstage, 10–15
    design of voltage-controlled current source, 15–16

260 Voltage Clamping

Voltage-clamp techniques (*continued*)
    membrane current measurement, 16–20
    modifying phase response of clamp amplifier, 20–23

Voltage clamping, single electrode
  clamp design, practical considerations in, 107–116
    antialiasing and output filters to minimize clamp noise, 110–112
    design of high-speed low-noise headstage, 107–108
    phase control and gain in clamp amplifier, 115–116
    selecting clamp sample-and-hold amplifier, 112–113
    selecting gain of controlled current source (CCS), 109–110
    setup procedure, 113–114
    smoothing measured current, 108–109
  examples of use of, 116–119
    clamping motoneurons in cat spinal cord with shielded microelectrode, 116–118
    clamping smooth muscle in guinea pig mucosa with high-resistance microelectrode, 118–119
  optimal clamping with, 95–120
  principles of operation, 96–107
    aliased noise and noise voltage index, 106–107
    clamp stability, 103
    microelectrode voltage artifact, 100–101
    selection of cycle period, duty cycle, and open-loop transconductance, 107
    steady-state clamp error, 101–103
    steady-state ripple in membrane potential, 105–106
    step response, 103–105

Voltage clamping, two electrode
  basic description, 54–56
  general considerations, 83–92
    background noise under voltage clamp, 86–90
    sensitivity to component variation, 83–85
    slowing the rise time of command potential, 85–86
    transfer-conductance voltage clamp, 90–92
  ideal voltage clamp: membrane capacitance as sole frequency-dependent component, 56–65
    equivalent circuit, 59–60
    membrane current responses, 60–64
    membrane voltage responses to changes in command potential, 56–59
  of cardiac Purkinje fibers, 172–178, 184–189
    clamp circuit, 184
    clamp circuit modifications, 186–188
    experimental guidelines, 184–189
    optimizing current microelectrodes, 184–186
    optimizing electrode spacings, 176–178
    testing the clamp, 188
  real voltage clamps, 65–83
    added phase lead to compensate amplifier input time constant, 70–73
    changes in bath potential, 81–83
    effects of series resistance, 73–77
    finite bandwidth of electronic amplifying circuit, 67–70
    interelectrode capacitance, 77–81
  speed of response of, 232–245
    clamp as linear feedback network, 233–234
    how to realize fast rise time, 245
    improving clamp response with current feedback, 242–245
    limitations because of amplifier bandwidth, 236–242
    two-time-constant control amplifier, 242
    wide-band clamp with different membrane elements, 234–236
  terminology and methods, 48–54
    cell model, 53–54
    Laplace transforms, 51
    potentials, currents, and null potentials, 51
    resistances and conductances, 51
    transfer functions, block diagrams, and stability, 51–53
  theory of electronic apparatus used in, 47–94

Whole-cell patch clamp: *see* Patch-clamp techniques